D0949820

p 5 3

Also available in the Bloomsbury Sigma series:

Sex on Earth by Jules Howard

p53

THE GENE THAT CRACKED THE CANCER CODE

Sue Armstrong

First published in 2014

Every effort has been made to trace or contact all copyright holders. The publishers would be pleased to rectify at the earliest opportunity any omissions or errors brought to their notice.

Bloomsbury Sigma is an imprint of Bloomsbury Publishing Plc
50 Bedford Square
London
WC1B 3DP

www.bloomsbury.com

Bloomsbury is a trademark of Bloomsbury Publishing Plc

Bloomsbury Publishing: London, New Delhi, New York and Sydney

A CIP catalogue record for this book is available from the British Library

ISBN (hardback) 978-1-4729-1051-6
ISBN (trade paperback) 978-1-4729-1320-3
ISBN (ebook) 978-1-4729-1053-0

10 9 8 7 6 5 4 3 2 1

Typeset by Mark Heslington Ltd, Scarborough
Printed and bound by CPI Group (UK) Ltd, Croydon, CR0 4YY

Bloomsbury Sigma, Book Two

For Struan, Isla, Louise and Fraser
who, I hope and trust, will reap the full rewards
of this mighty endeavour in cancer research

Contents

Preface

> *Where do we* [scientists] *get our ideas, our inspiration for solving problems? It's the same place a composer gets an idea for a piece of music, or a painter gets an idea for a painting. It comes out of somewhere that you don't know. It's the same flash of inspiration, and it's associated with the same colour – and the same glory, for want of a better word.*
>
> Gerard Evan

***

Luana Locke is a vivacious woman on the threshold of middle age, with a round pretty face, large hazel eyes and a tumble of wavy dark hair to her shoulders. There is such an air of robust good health about her as she sits talking over a frothy cappuccino in a busy Toronto café that if I didn't already know something of her story I would never suspect that her life has been dogged by sickness, heartache and loss. A survivor of cancer, Luana already had long experience of the disease by the time of her own diagnosis at the age of 24. When she was just three years old, her sister Manuela, aged nine, died of a brain tumour. Of that period, Luana can only really remember being left frequently with relatives as her parents visited the hospital, her mother's terrible grief when Manuela finally died and her own helpless desire to see her smile again.

'I remember one time giving her a tissue and being upset because she blew her nose. I thought, "No! I gave it to you to wipe your tears." I just remember her being really really sad.' The little girl did not know then that her mother Giulietta was also sick. She and Luana's father Franco, a tile-setter by trade, had emigrated from Italy to Canada a few years earlier for what they intended to be a short spell taking advantage of the high demand for skilled craftsmen. They were planning to go home to Italy – drawn partly by the fact that Luana's Aunt Rina, her mother's twin sister, had recently died of breast cancer at the age of 29, leaving four small children. But their plans were scuppered when their own daughter was diagnosed with a brain tumour, and then Luana's mother began her own treatment – chemotherapy and radiation for breast cancer.

'She had a mastectomy, and I do have some very clear images of that time,' says Locke. 'I remember watching my mother as she was getting dressed, putting on her make-up, fixing her hair and slipping her prosthesis under her shirt.' All this seemed normal, just part of life, to the small girl, until her mother's cancer, which had responded to the initial treatment, returned and spread to her bones and killed her. Luana was six years old, and she speaks poignantly of the heartache and emptiness that filled their home, reduced now to her father, her brother David and herself, and of the fear that gripped her in bed at night that this monster might strike again and take more of her loved ones.

In fact the next person to be diagnosed was Locke herself. She was 24 years old and eight months pregnant when she noticed a tiny scab on her nipple. It would slough off from time to time leaving a small, weeping sore that would scab over again, but never seemed to heal. Keen to breast-feed her baby when it arrived and anxious therefore to clear up the spot, she eventually went to her doctor, who gave her some ointment. No one she saw as part of her routine maternity care seemed concerned, but when the spot appeared to be getting bigger despite a variety of ointments, her doctor referred her to a dermatologist, Donna McRitchie, who decided to take a biopsy.

'Dr McRitchie said, "It will probably take about a week to get the results back, and we'll call you,"' says Locke, looking back across the years. 'I remember going and sitting in my car and starting to cry. Even though I didn't feel anything

– she put a needle in to freeze the area – I remember hearing the clipping of the scissors and I knew that she was cutting tissue, and it really affected me in a weird kind of way. So I got into the car and cried . . . But then I stopped and I was so angry with myself. I thought: You baby! Think what your mother went through; she had her breast *removed*, for the love of God, and here you are weeping like a baby because they took a little piece of skin off your breast. Like, grow up and get over it! I was just kind of scolding myself, right? Snap out of it, off you go.'

Luana was not ready to admit her fears even to herself. A naturally optimistic person, she hung on to the belief that lightning could not strike again in the same spot, and whenever a morbid thought entered her mind she would squash it, telling herself, 'That's ridiculous; who's ever heard of breast cancer manifesting itself that way? There's no lump; there's nothing there; you're 24 years old . . . Gosh, I scolded myself a lot in those days!' she laughs. But the results came back unusually fast, and they were serious: Luana had Paget's disease, a type of breast cancer that is often mistaken for eczema until the tumour growing secretively below the sore is advanced. Given her family history, her oncologist recommended the most radical treatment, mastectomy.

The diagnosis was a shock, but one of Luana's first concerns was for her father. 'Telling him was one of the hardest things I ever had to do,' she comments quietly, looking down and stirring the froth in her coffee cup. 'He'd already been through so much and it was just a journey I didn't want him to have to travel again.' In the event, Franco put on a brave face, urging his daughter and her distressed husband Paul not to look back to her mother's and sister's experience nearly 20 years earlier. '"Medical science has come such a long way since then," he said.'

Within weeks of Luana's diagnosis, her baby, a boy they named Lucas, was delivered by Caesarean section; her mastectomy was performed a few days later. Examination of the tissue removed at surgery revealed a highly aggressive tumour, and Luana subsequently had her other breast removed as a precautionary measure. Her surgeons found a pre-cancerous lesion here too. During all this, Luana coped with her own disease, with the painful memories it inevitably stirred and with the dreadful uncertainties of the future by focusing on her new baby. 'It was all about, okay, I know I have to survive, so I'll do whatever it takes . . . I really just poured myself into *nesting*,' she says. 'The worst thing that I allowed myself to contemplate was not my death – that was too large, too big an issue for me to face, too big a fear, I guess . . . the farthest I allowed myself to go in terms of my biggest fear was losing my hair. I've always had long hair, and so I allowed myself to go there, but I didn't allow myself to think of death . . . Or not being here for my child, no, not at all.' She shakes her head.

Luana is now 41, and she has not had a recurrence of her cancer, though in the intervening years four other members of her immediate family in Italy and America have had tumours removed, and her brother David's little boy Marco died of cancer, aged five. Though no one knew it until relatively recently, the root of the family's problem is a mutation in a gene that goes by the prosaic name of p53 – bestowed on it simply because it makes a protein with a molecular weight of 53 kilodaltons.

When it was discovered in 1979, the scientists involved had little idea of the huge significance of their finding: p53 has gradually revealed itself to be one of the most important players in the drama that is cancer – a master switch in our cells whose main function is to prevent tumours arising when their DNA is damaged. It has become the most studied single gene in the history of molecular biology, generating over 70,000 research papers to date and spawning a community of researchers that, notwithstanding the customary competition between scientists, is unusually collaborative. Every two years they come together from around the world for a scientific meeting, a few days of bracing, esoteric debate which adds new bits to the mighty jigsaw and fits some old ones into the existing picture.

p53 is the most commonly mutated gene in human cancer. This means that the gene is damaged and the information it carries is altered, in much the same way as a CD or computer file can be corrupted and its information scrambled. In those cases where the gene is not mutated, typically some other abnormal event in the cell is preventing it from functioning as it should. 'There are lots of other genes you see mutant in the various tumour types,' comments Bert Vogelstein of Johns Hopkins University in Baltimore, Maryland, 'but p53 is one of the few that goes across the board. It's unique in that it's a common denominator of cancers.'

Vogelstein, who was born and brought up in the shadow of Johns Hopkins in the 1940s and went to medical school there, has been involved with p53 since its earliest days. His lab, housed today in a tall modern building of glass and light which looks out over Baltimore and down onto the warm red bricks of the old hospital, has provided some of the most important insights into the workings of the gene. 'I think you could safely say that it's impossible – or very difficult – to get a malignant tumour without the activity of p53 being disrupted.'

***

I first heard of p53 in 1996 when I was newly returned to Scotland from seven years in South Africa, where I had been reporting for *New Scientist* magazine and BBC Radio, and chronicling the ravages of AIDS across the continent for the World Health Organization. Now I was looking around for interesting science stories in Scotland, and I found myself in the lab of David Lane, one of the four discoverers of p53. Intrigued by what I learnt, I got a commission from the BBC to make a radio documentary and flew off to Crete to join the p53 community at one of their biannual workshops.

The setting was a conference centre perched on the lip of a sweeping bay with views out over private white sands and the sparkling ocean. At the end of the first session I sat stunned – the scientists could have been speaking Greek for all I had understood.

Sitting next to me at the noisy dinner table later that evening was Peter Hall, small, bright, mischievous and a very good teacher, a scientist from Dundee with whom I'd worked before on p53 stories. Leaning towards me, he whispered, 'Don't panic. This is what you need to know to get a handle on the debate ...' He pointed me towards the most interesting presentations in p53 research and the people I must corner with my microphone. I began to relax and returned from Crete four days later with enough material and a good enough story for not just one but two radio documentaries that looked at the basic science of this gene and at the promise it held for new kinds of cancer therapy.

That was 1998, and over the following years I have revisited the story of p53 whenever new bits of the jigsaw puzzle have piqued my interest, such as the chance discovery, when a genetically engineered mouse experiment went dramatically wrong, of an intimate connection between cancer and ageing; and of the role of p53 in nailing the tobacco industry by furnishing unequivocal proof that smoking is a direct cause of cancer. I have watched as morale among p53 researchers has waxed and waned in the light of new information that carries them forward on a wave of discovery and enlightenment one moment, then mires them once more in the fog of complexity.

Over the years I began to realise that this was too good a story to leave to the passionless pages of academic journals, where it's hard for the layperson to grasp the significance of even some of the most momentous discoveries; the idea of writing a book about the gene took root. It's not a straightforward story, because nothing in science ever is. Knowledge advances as much through negative results and thwarted hypotheses as it does by theories that prove to be correct. It takes open minds and intellectual courage to recognise the absence of proof – or the totally counterintuitive outcome of an experiment – as offering important insights in their own right. In their quest to understand the workings of cancer, p53 researchers have been as much influenced by dogma and the power of paradigms as have scientists in any other field.

'Science without storytelling,' wrote the American astrophysicist Janna Levin in a commentary for *New Scientist,* 'collapses to a set of equations or a ledger full of data.' My aim here is to stand clear of those ledgers full of data as far as possible and tell the story of some of the curious, obsessive, competitive minds that filled them, thereby helping to unravel the deepest mysteries of cancer.

A NOTE FROM THE AUTHOR

I have tried hard to avoid jargon as far as possible. But there is one expression I am loath to replace with something more readily understood: that expression is 'wild type' in reference to the status of the p53 gene. Essentially, the 'wild-type' gene means the 'normal' gene that functions as nature intended, as opposed to the 'mutant' gene whose behaviour is aberrant. 'Wild type' is a term so widely used by the biology community, and by my interviewees – and so much more vivid than 'normal' – that I have decided not to substitute it. I trust my readers will forgive me.

CHAPTER ONE

Flesh of our Own Flesh

In which we learn that cancer is more than 200 different diseases, but they all share some common characteristics – the most important being that, if p53 is functioning properly, a cell cannot turn malignant.

***

Tumours destroy man in a unique and appalling way, as flesh of his own flesh which has somehow been rendered proliferative, rampant, predatory and ungovernable.

Peyton Rous

'The question that's obsessed me for the whole of my career is: why is cancer so rare?' Gerard Evan, a professor of molecular biology at the University of California, San Francisco, and Cambridge, England, pauses to let his comment sink in. He knows it will startle me, for the statistics most commonly quoted in the media paint a bleak picture: that one in three of us will be diagnosed with cancer at some point in our lives and one in four of us will die of the disease. But Evan, talking to me in his office in the Sanger Building in a leafy corner of Cambridge about his years of research at the most fundamental level of the genes, is looking at cancer from the viewpoint of the cells, not of the whole human being. It takes just one rogue cell which has lost its normal regulatory machinery and run haywire to trigger cancer, yet billions upon billions of cells in our bodies that are growing and replicating themselves all the time do so typically for 50, 60 years or more without producing a tumour. And in two in three of us they never do. 'I mean, if you were doing the lottery you'd never gamble on this!' continues Evan. 'Cancers do arise, but clearly we've evolved amazingly elaborate and effective mechanisms to restrict the spontaneous evolution of autonomous cells within our bodies. And even though we bomb ourselves with mutagens and carcinogens and do all sorts of things we shouldn't do, still most people die of heart disease; they don't die of cancer.'

A measure of just how resistant our cells are to corruption is the fact that a goodly chunk of our DNA – nature's instruction manual for building our bodies – can be traced back to the original single-celled organism known as the 'last universal common ancestor' of all life on earth (often referred to by the acronym LUCA), whose existence was first proposed by Charles Darwin in his book *On the Origin of Species,* published in 1859. In other words, some of our genes are more than 3.5 million years old and have been passed down faithfully from one generation to the next over unimaginable eons of time.

The term 'cancer' represents not one but a collection of around 200 different diseases which share this common characteristic: they all originate from a single cell that has become corrupted. The great majority of cancers – well over 80 per cent – are carcinomas, which means they are in the epithelial cells that form the outer membranes of all the organs, tubes and cavities in our bodies, and include our skin. The connective tissue, which provides the structural framework for our bodies, and support and packaging for the other tissues and organs – it includes, for example, bone, cartilage, fibrous tissue such as tendons and ligaments, collagen and fatty tissue – appears extremely resistant to turning malignant. Sarcomas, which are cancers of the connective tissue, account for only about one in a hundred cases.

No one yet knows the reason for this bias, though speculation is intense. Could it be that epithelial cells tend to divide more often than connective tissue cells and the opportunity for mutation is much greater? Our skin, for instance, has an intense programme of self-renewal with cells at the base layer dividing and undergoing processes of differentiation and maturation as they push up towards the surface, where they are eventually sloughed off (that's what causes the tidemark around the bathtub). The lining of the gut, too, is constantly renewing itself, and the sloughed cells are excreted. However, an argument against high rates of proliferation being the main reason why epithelial cells are at greatest risk of malignancy is the fact that some of the most cancer-prone epithelial cells are not ones that divide most frequently. Some suggest that it is because epithelial cells are a first line of defence against the outside world and are more likely to come into contact with cancer-causing agents. But this argument too has weaknesses, since epithelial and connective tissue cells are equally exposed to carcinogens in some organs, notably the prostate, yet the epithelial cells are the more vulnerable.

Looking for answers to this conundrum, one lab took samples of healthy breast tissue, teased apart connective tissue cells from epithelial cells and watched what happened when they attacked them with chemical carcinogens in their Petri dishes. To their surprise, they saw that the two cell types reacted completely differently, though they still don't know exactly how or why. That's the Holy Grail, as it might point to chinks in cancer's armour as targets for new drugs.

Tumours typically arise from the pool of stem cells in a tissue that are responsible for the repair and replacement of cells as part of the routine maintenance of our bodies. It can take years, even decades, for a rogue cell to grow into a tumour that is detectable. This is because it depends on progressive breakdown of the cellular machinery through the mutation and/or loss of crucial genes that regulate growth, replication, repair and timely death of cells – mutations that occur independently and, crucially, don't result in the cell being eliminated, which is the normal fate of damaged cells. The growing tumour is parasitic: it competes with the normal cells around it for nutrients and oxygen, and it can't grow much beyond 1–2mm (1/25th–1/12th of an inch) in diameter unless it develops its own blood supply.

What distinguishes a malignant tumour from a benign one is the former's ability to spread – to send out microscopic shoots that penetrate the walls and invade neighbouring tissue, and to seed itself in distant sites from breakaway cells carried in the bloodstream or lymph system. Blood-borne dissemination is particularly efficient at spreading cancer, with the blood depositing its cargo of delinquent cells along natural drainage sites, most commonly the liver and lungs.

Our understanding of the mechanics of cancer has advanced at revolutionary speed in the last 40 years, as one technological breakthrough after another in molecular biology has enhanced scientists' ability to explore the workings of the cells – the building blocks of all life on earth. But the sheer volume of data churned out has threatened at times to overwhelm the cancer research community. Robert (Bob) Weinberg has been involved since the early 1960s and has played a big part in the revolution. 'The plethora of information is just overwhelming,' he told an audience of mostly fellow scientists who had gathered in a lecture theatre at Massachusetts Institute of Technology, MIT, to hear him speak of his life in science and of the personal experiences that had formed him.

'When I was a graduate student there were two journals one paid any attention to: the *Journal of Molecular Biology* and *PNAS* (*Proceedings of the National Academy of Sciences*). That was it. Today?' Weinberg shrugged mightily and spread his hands. 'More than the stars in the sky. I think PubMed* now has 12 or 15 million papers in it, and the only way I can deal with this is to continually ask people who have distilled information in their own minds how this or that problem is evolving.' Laboratory scientists today can generate important data up to 10,000 times faster than he could when he started out in cancer research, Weinberg told his audience.

---

* A free database of references to papers on life sciences and biomedical topics set up in 1996]

In his own lab at MIT, where he has spent most of his working life, Weinberg puts pressure on people to draw lessons and develop ideas from what they observe, not simply accumulate data. It's little surprise, then, that he should have become preoccupied with bringing some order to the explosion of information about cancer that in many ways mirrors the disease's own chaotic growth. Attending a conference in Hawaii in 1998, Weinberg took a walk down to the mouth of a volcano with fellow scientist Doug Hanahan, who had also caught the molecular-biology bug at MIT as an undergraduate. Mulling over their common frustration as they walked, the two conceived the idea of writing a review that would seek to clarify what Weinberg calls the 'take-home lessons' of research.

'Cancer research as a field was a very broad and disparate collection of findings, and we thought there might be some underlying principles through which we could organise all these disparate ideas,' he said. 'We came up with the notion that there were six properties of cancer cells that were shared in common with virtually all cancer cells and that defined the state of cancerous growth.' 'The Hallmarks of Cancer' was published in 2000 and far from disappearing 'like a stone thrown into a quiet pond', as Hanahan and Weinberg had predicted, knowing how quickly most journal articles are read and forgotten, their paper has become the descriptive cornerstone of cancer biology and a clear framework into which new pieces of the jigsaw can be slotted. The six characteristics they identified as being common to virtually every cancerous cell are that, in lay terms:

- the forces pushing them to grow and divide come from within the corrupted cell itself, rather than being signals from outside;
- cancer cells are insensitive to forces that normally stop cell division at appropriate times;
- they are resistant to being killed by the mechanisms that normally remove corrupted cells;

- they are immortal, meaning they can divide indefinitely, whereas normal cells have a finite number of divisions controlled by an internal 'clock' before they stop dividing, become senescent and eventually die off;
- they develop and maintain their own blood supply;
- they can spread to other organs and tissues and set up satellite colonies, or metastases.

In 2011 the two scientists updated and refined their 'Hallmarks' paper, adding further general principles, including the fact that the metabolism in cancer cells – particularly the way they use glucose to provide energy – tends to be abnormal; and that they are able to evade detection and destruction by the body's immune system.

Crucially for the story I'm telling here, p53 plays a role in all these traits. 'As I read the paper by Hanahan and Weinberg, I said, "This is conceptually brilliant!"' comments Pierre Hainaut, who spent many years investigating cancer genetics at the World Health Organization's International Agency for Research on Cancer, IARC, in Lyon, France. 'But then I thought: where is the unity? Clearly it must be a more coherent programme than just a succession of boxes. What is holding it together? And then I realised: my goodness, it's p53!

'There are many genes that have a mechanistic role in one hallmark trait or another, and this will spill over to two or three hallmarks. But p53 is the one that links all the hallmarks together. This means that from a molecular viewpoint there is one basic condition to get a cancer: p53 must be switched off. If p53 is on, and hence functioning properly, cancer will not develop.'

Hainaut – a tall, rangy Belgian with a crooked smile, boyish enthusiasm and an earnest expression behind black-rimmed specs – has been particularly intrigued by the gene's multiple roles in the activities of the cell. It's an interest that takes him frequently out of his lab and into the wider world, to investigate the connection between mouldy peanuts and liver cancer in the Gambia, to meet families with a hereditary cancer disposition in southern Brazil, and to many other countries, from China to Iran, in pursuit of insights into the workings of p53 in our everyday lives. 'There are many, many ways to lose the function of p53,' he continues. 'Mutation is a very common one; loss of one copy of the gene is another; but there are also other ways, such as switching it off, degrading it, putting it off-site and so on. But I repeat: if the cell is retaining a perfectly intact and fully reactive p53 function, it will not give rise to cancer.'

AN ANCIENT MALADY

Cancer is a disease as old as humankind. It is mentioned in the earliest medical texts in existence, a collection of papyri from ancient Egypt dating from 3000–1500 BC. Actual specimens of human tumours have been found in the remains of a female skull from the Bronze Age, dating between 1900 and 1600 BC, and in the mummified remains of ancient Egyptians and Peruvian Incas. In 1932, the palaeoanthropologist Louis Leakey, working in the Rift Valley of East Africa, found evidence suggestive of bone tumours in the fossilised remains of one of our hominid ancestors, *Homo erectus*, who roamed the African savanna between 1.3 and 1.8 million years ago.

In fact, cancer has probably been around since LUCA first gave rise to multi-celled creatures. In 2003 a team from Northeastern Ohio Universities College of Medicine, led by radiologist Bruce Rothschild, travelled around the museums of North America scanning the bones of 700 dinosaur exhibits. They found evidence of tumours in 29 bone samples from duck-billed dinosaurs called hadrosaurs from the Cretaceous period some 70 million years ago. And evidence of tumours has been found also in the bones of dinosaurs from the Jurassic period between 199 and 145 million years ago.

Hippocrates, living in ancient Greece around 460 BC, was the first person to recognise the difference between benign tumours that don't invade surrounding tissue or spread to other parts of the body, and malignant tumours that do. The blood vessels branching out from the fleshy growths he found in his patients so reminded him of the claws of a crab that he gave this mysterious disease the name *karkinos,* the Greek word for crab, which has translated into English as carcinoma. Hippocrates and his contemporary physicians believed cancer was a side effect of melancholia. And up to the Middle Ages and beyond, medics and patients alike reckoned the causes were supernatural and related to demons and sin and the accumulation of black bile.

This menacing theory of cancer prevailed for nearly 2,000 years before it was exploded by Andreas Versalius, a Flemish doctor and anatomist working in Padua, Italy, in the early 16th century. Versalius performed post-mortems on his patients, as well as dissecting the corpses of executed criminals supplied to him by a judge in Padua fascinated by his work: black bile, he announced, was nowhere to be found in the human body, diseased or healthy.

But it was another two centuries and more before anyone suggested that agents in our environment might be playing a part in the development of tumours. In 1761 John Hill, a London physician and botanist, produced a paper, 'Caution Against Immoderate Use of Snuff', in which he described patients with tumours of the nasal passages as a consequence of sniffing tobacco. And in 1775 an English surgeon, Percivall Pott, reported a number of cases of cancer of the scrotum in unusually young men whose only link was that they had been chimney sweeps as small boys and were likely to have gathered soot in the nooks and crannies of their bodies as they squeezed themselves up the narrow flues of homes and factories in Georgian Britain – a practice that lasted for two centuries and frequently involved children as young as four years old. In 1779, the world's first cancer hospital was set up in Reims, France – at a fair distance from the city because people feared the disease was contagious.

The foundation of our modern understanding of cancer as a disease of the cells was laid in the mid-19th century by Rudolf Virchow, a German doctor born into a farming family, who won a scholarship to study medicine and chemistry at the Prussian Military Academy. Often referred to as the father of modern pathology, Virchow was much less interested in his suffering patients than in what they suffered from – the mechanics of disease – and preferred to spend his time in the lab poring over his microscope and doing animal experiments than visiting the sick. The idea that living cells arise from other living cells through division had been around for many decades but had been almost universally rejected, perhaps because it offended religious sensibilities about creation – in those days people really believed that maggots appeared spontaneously in rotting meat.

It was not until the strong and independent-minded Virchow, who was active in politics as well as in science and medicine, published his own observations of cell division and coined the phrase *omnis cellula e cellula* – which translates roughly as 'all cells arise from other cells' in a continuous process of generation – that the idea finally caught on. But it was the advent of molecular biology in the mid-20th century that has allowed scientists to peer ever deeper into the cell – to study the machinery of life itself in the DNA – and to begin to crack the code of cancer.

CHAPTER TWO

The Enemy Within

In which we hear a) about a virus that causes cancer in chickens that can be passed on to other birds in the DNA and b) of the discovery of the first genes responsible for driving cancer – the so-called oncogenes.

***

Now more ambitious questions arose . . . Might all cancers arise from the wayward action of genes? Can the complexities of human cancer be reduced to the chemical vocabulary of DNA?

Michael Bishop

At the cutting edge of research in the mid-1900s was the idea – incredibly controversial at the time because there was no direct evidence for it – that viruses could cause cancer in humans. The central figure in this story is Peyton Rous, born a full century earlier in Texas, who studied medicine at Johns Hopkins in Baltimore and was very nearly lost to science before he began. While still a student, Rous contracted tuberculosis from a cadaver he was dissecting when he cut his finger on a tuberculous bone. He had surgery to remove infected lymph glands, and was sent home to recuperate under the big skies and fresh air of a Texas ranch. A year spent rounding up cattle on horseback and sleeping out under the cold stars on the range with the other cowboys – a profound experience from which he drew pleasure for the rest of his life – restored him to health and he returned to Johns Hopkins Medical School.

Rous qualified in 1905, but during his first year on the wards he, like Virchow before him, decided he was not cut out to care for the sick, and he retreated from the front line to the pathology lab to study disease. By 1909, he found himself in charge of the cancer laboratory at the Rockefeller Institute of Medical Research – a post vacated by the institute's director, Simon Flexner, who wanted to turn his attention to the more pressing problem, as he saw it, of polio, which was crippling millions of American children.

A trawl through the history of cancer research draws one up sharp: almost everything we know today about cancer – as a disease of the cells and of the genes – was suggested by someone way back before scientists had any way of testing their ideas, and who is often forgotten by those who later reveal them as facts when the world is more ready to listen. However, in 1909, precious little had been established about how cancer works and, apart from surgery and the newly discovered but not yet widely available use of X-ray radiation, there was no treatment. Much of the research effort at that time went into expanding the insights of people like John Hill and Percivall Pott that chemicals cause cancer, and identifying these carcinogenic agents.

But Rous was interested in exploring another idea: whether or not viruses could cause malignancy. He could hardly have asked a more difficult question to investigate, for at that time viruses were more of a concept than a reality, known by their footprints only. Viruses had been 'discovered' in the 1890s when scientists working on infectious agents developed filters that could block the passage of bacteria. When they found that a filtered solution from which all pathogens then known had been removed remained infectious, they scratched their heads in perplexity and concluded that whatever was causing the infection must be a chemical. They then gave it the Latin name for poison: 'virus'.

By the time Rous was asking his questions, the idea of viruses as living entities was more or less accepted by the scientific world, and a few that caused diseases in plants had already been identified. But viruses could neither be seen, nor cultured, nor caught in filters. They could be identified as the infectious agent only by exclusion – when other, larger pathogens had been caught in the net.* A virus that causes leukaemia in chickens had already been discovered, in 1908, but the finding caused hardly a stir in the cancer community because leukaemia was not then recognised as a malignant disease. Rous, however, was intrigued and in his search for a cancer-causing virus he turned his attention to barnyard fowl. Very quickly he struck lucky. In 1910 he discovered a sarcoma – a cancer of the connective tissue – in chickens that could be induced in healthy birds by injecting a filtered extract of the tumour from a sick bird into a healthy one, where in time it produced exactly the same type of tumour as the original. What's more, his experiment worked over and over again, and he believed he could detect signs of the virus in the tumour cells.

Rous published his findings in 1911 but they were roughly and widely dismissed, as he told his audience at the Swedish Academy on receiving the Nobel Prize for Medicine decades later in 1966. 'Numerous workers had already tried by then to get extraneous causes from transplanted mouse and rat tumours, but the transferred cells had held their secret close. Hence the findings with the sarcoma were met with downright disbelief.' This was despite the fact that the experiment with barnyard fowl had been repeated successfully on several more occasions, and a virus found each time. 'Not until after some 15 years of disputation amongst oncologists were the findings with chickens deemed valid – and then they were relegated to a category distinct from that of mammals because from them no viruses could be obtained.'

In the event, Rous's work with chicken viruses was to spawn one of the most exciting and productive fields ever in cancer research. But he died in 1970, just too early to

* When the electron microscope was developed in 1931, virus particles could be seen for the first time.

see it truly bear fruit. 'Tumours are the most concrete and formidable of human maladies, yet despite more than 70 years of experimental study they remain the least understood,' he told his Nobel audience, musing a little later, 'We term the lawless cells neoplastic because they form new tissue, and the growth itself a neoplasm; but on looking into medical dictionaries, hoping for more information, we are told, in effect, that *neoplastic* means "of or pertaining to a neoplasm", and turning to *neoplasm* learn that it is "a growth which consists of neoplastic cells". Ignorance could scarcely be more stark.'

After experiencing the same failure as others to repeat his chicken results with rats and mice, Rous quit virus research for more obviously rewarding fields of pathology. It fell to others to tease out what Rous sarcoma virus, RSV, as it had been named, was doing in cells to make them malignant, and to see what lessons this might hold for understanding the mechanics of cancer in humans.

A DISEASE OF THE GENES?

Notable among these others are Michael Bishop and Harold Varmus who, working together at the University of California at San Francisco in the early 1970s, made such momentous discoveries with RSV that they too won the Nobel Prize for Medicine, in 1989. Bishop had begun work with the virus in 1968, just two years after Rous's Nobel award, and says of that event that it 'dramatised the great mystery of how RSV might cause cancer. It was a mystery whose solution lay in genetics.'

This was an assessment shared by Varmus, then doing postgraduate training in medical research on the other side of the country at the National Institutes of Health, NIH, in Bethesda, Maryland. Varmus had become excited at the potential for approaching the mind-boggling complexity of human genetics – particularly in relation to disease – through the study of much simpler organisms. 'From some dilatory reading in the early 1960s, I knew enough about viruses and their association with tumours in animals to understand that they might provide a relatively simple entry into a problem as complex as cancer,' he wrote in his autobiographical account of his work. 'In fact, for anyone interested in the genetic basis of cancer, viruses seemed to be the only game in town.' The little scraps of life contain around five to ten genes all told, compared with 20,500 or so in our cells.

In the summer of 1969, Varmus and his journalist wife Connie combined a backpacking trip to California with the search for opportunities to study viruses on the West Coast. Visiting Mike Bishop in his lab at UCSF, he found a fellow book addict with wide tastes in literature, and a keen writer also. Bishop, too, had been ambivalent about becoming a doctor, but had discovered almost by chance the thrill of laboratory science and not looked back. In Bishop, Varmus had discovered a kindred spirit and he agreed to join him the following year. 'Harold's arrival changed my life and career,' Bishop recalled during his Nobel address. 'Our relationship evolved rapidly to one of co-equals and the result was surely greater than the sum of the two parts.'

When they started their work together the two scientists were somewhat out on a limb, for in 1970 many of their peers were still sceptical, even frankly disbelieving, of the theory that cancer is a disease of the genes, since there was no direct evidence for it. But that very summer, a young postgraduate student from California named Steve Martin appeared at the Gordon Conference, an annual event that brings together scientists at the cutting edge of their field internationally to brainstorm ideas in the informal setting of an old boarding school in Tilton, New Hampshire. The topic at that year's conference was animal cells and viruses, and Martin – a 'bookish' young man, 'with dark curls, a cherubic face and an enthusiastic manner', according to Varmus – had come to tell his colleagues that he had managed to isolate the gene in the Rous sarcoma virus responsible for turning infected cells delinquent, and to explain how he had done so. The gene – soon given the name Src (pronounced 'sark') after the type of tumour it causes – was the first example of what became known as 'oncogenes'.

Derived from the Greek word *onkos* meaning 'mass', and describing a gene that can transform a normal cell into a tumour cell, oncogenes are at the very heart of the story of p53. What earned Bishop and Varmus their Nobel Prize was the discovery in 1974 that the *normal* cells of uninfected chickens have copies of a gene almost identical to the Src found in the virus. What is more, the two scientists soon found that many other bird species – including ducks, turkeys, quails and even an emu from the Sacramento Zoo – did too. In due course, Src-like genes would be detected in fruit flies and worms and many species of mammals, indicating that this was a gene that could be traced back through eons of evolutionary history and that must therefore have an essential role to play in the cell.

But in 1974, the evidence from bird species alone was enough for Bishop, Varmus and their team to suggest a revolutionary idea: that, rather than being the carrier of an alien gene with which it corrupts host cells, the virus had, some time in the course of its evolution, picked up the gene from its chicken host and incorporated it into its own genome – a process that caused the gene to become dangerous to its original bird host. Could it be, they speculated further, that there are other genes – perhaps many of them – in normal animal cells that are capable of becoming oncogenes when picked up and spread around by viruses?

ARE WE CREATING FRANKENSTEIN SPECIES?

Attempting to answer such questions was extremely painstaking work with the tools then available – and the technical difficulties were compounded by an atmosphere of high anxiety and ethical soul-searching among scientists at the time. Genetic engineering was in its infancy. It had begun in the 1960s with work on a class of viruses that infect only bacteria and nothing else. Scientists had discovered that these viruses – known as 'bacteriophages' (often contracted simply to 'phages') – have a habit of taking up genes from one bacterium they have infected and transferring them to their next bacterial victims; researchers had begun to harness this trait to investigate the activity of genes from a wider variety of organisms.

The bacterium most commonly used as a vehicle for the scientists' experiments was *Escherichia coli* (E. coli), because it readily takes up new genes, is tough and breeds fast, thus offering them swift results. But as the technology developed and scientists became adept at splicing genes – chopping out bits of DNA from one genome and stitching them into another to produce what is known as 'recombinant DNA' – some among them grew concerned.

They were no longer working only with bacteriophages, but with animal viruses also. And, as is the custom, there was a good deal of sharing of hybrids between labs worldwide, so the novel scraps of life were being ever more widely distributed. But were these curious scientists – with their minds focused on cracking the secrets of the cell – wandering in where angels feared to tread? While many were hugely excited by the potential of the new gene-splicing tools to deliver better ways of treating disease and improving food crops, others feared the potential for catastrophe, wrote M J Peterson, in a case study for the Science, Technology and Society Initiative of the University of Massachusetts. 'They believed that the probability of unintentionally creating dangerous organisms – virulent "super" versions of disease-causing viruses or bacteria, strange and invasive life forms that push others out of habitats – because the ways genes and gene sequences interact were not well understood, was too high to be ignored.'

Researchers were using a strain of E. coli known as K12 that had evolved to a stage where it could no longer live outside of cell cultures in test tubes and Petri dishes. However, the fear was that, if it ever escaped the lab, K12 might combine with other strains of E. coli – a bacterium widespread in nature and living harmoniously for the most part with us and other animals. Most concerned about the Pandora's Box they might be opening were the microbiologists, who fretted that those without their specialist training in the life habits of micro-organisms – which category included many researchers who had eagerly taken up the new tools – did not fully appreciate the risks. Often they would tip biological waste that contained bacteria down the drain at the end of experiments, or draw up solutions by mouth with a pipette. As the critics pointed out tartly in a report on the possible hazards some time later, 'A micro-organism is not simply a "warm body" to house a recombinant DNA molecule of interest.' In 1974 a number of leading scientists stopped their work on recombinant DNA pending a formal debate on the way forward for laboratories using this technology.

The following year the intense soul-searching among scientists, and the equally volatile debate that had begun in the world's media, culminated in an international conference held at the Asilomar Center, a magnificent old lodge built of warm local wood and stone overlooking the Pacific near Monterey, California. Writing for *Science* magazine in 2000 on the 25th anniversary of the Asilomar Conference, journalist Marcia Barinaga called it 'the Woodstock of molecular biology: a defining moment for a generation, an unforgettable experience, a milestone in the history of science and society'. Looking back across the years, David Baltimore, who won the Nobel Prize for Medicine in 1975 for his work with viruses and was one of the organisers of the conference, said, 'Recombinant DNA was the most monumental power ever handed to us. The moment you heard you could do this, the imagination went wild.'

In fact, so exciting was it, and so potentially scary, that the attempt to reach consensus on the way forward among the disparate group of 133 scientists gathered at Asilomar – debating under the watchful eyes and listening ears of 16 journalists and four lawyers – was extremely difficult. What eased the process was the decision to divide the types of experiments using recombinant DNA into several categories – depending on whether they involved organisms or fragments of DNA known to cause disease or pose other dangers, or used materials considered harmless – and making recommendations about how to proceed under different scenarios. These included taking measures to disarm living organisms used in experiments so that they could not interbreed nor survive outside of tissue cultures; and adopting specific safety measures in the design of labs. It fell to national governments to turn the recommendations of the Asilomar Conference into useable guidelines, and by 1976 scientists were able to resume their experiments with recombinant DNA, more or less reassured that they were not about to unleash Frankenstein species upon the world.

HOW ONCOGENES TURN NASTY

A question burning in the minds of researchers ever since Bishop and Varmus's discovery of the Src oncogene in chickens and other birds was: could these cellular oncogenes cause cancer without first being captured by a virus? They found that indeed they could. The first evidence came from labs, including Bishop and Varmus's own, which were investigating the action of viruses that infect animals and cause cancer, but that don't actually possess oncogenes. What the researchers found was that these viruses 'hit' the DNA of the host cell in animals, corrupting its information and diverting it from its normal function as part of the growth and repair machinery to drive a cancer programme instead.

Further evidence came from Bob Weinberg's lab at MIT which was looking not at viruses but at how certain chemicals can cause cells to become cancerous. Weinberg was intrigued by Bishop and Varmus's work and it led him to wonder if chemical carcinogens worked in the same way – by corrupting would-be oncogenes (so-called 'proto-oncogenes') and turning them nasty. At that time, even though the idea that mutant genes were what drove cancer had been around for some time, there was still very little hard evidence, lots to test and many doubters. Weinberg himself was not yet fully convinced.

To investigate his theory Weinberg's team took mouse cells which they treated with chemicals to make them turn cancerous. Then they extracted the pure DNA from these cells, and inserted it into normal mouse cells in Petri dishes in the lab. Sure enough, the normal cells turned cancerous too – indicating that the agent that was causing the cells to turn cancerous was carried in the DNA, in the genes, though they didn't yet know which gene or genes was the culprit. This was the first evidence that would-be oncogenes did not need the action of viruses to turn nasty; mutations caused by chemicals could have the same effect.

However, these exciting results were somewhat overshadowed by a scandal involving a scientist in Canada who had been doing research along the same lines. 'This person was invited to give a seminar at Harvard,' Bob Weinberg told his audience at MIT. 'I attended it and I was both extraordinarily impressed and extraordinarily depressed. The work that was shown was so vast that it was far beyond anything my laboratory could ever do. It was *beautiful* data, and it indicated to me that this guy had already done more than my own laboratory could do for the next 10, 20 years – just a *vast* amount of data! And I went up to this guy and I said, "You know, we're beginning to get results just like these." And he said, "*Really?*" and he was very excited.

'Now I will tell you that usually when you have somebody coming up to you after a seminar and telling you that they're getting exactly the same results as you have recently been getting, you have mixed feelings,' Weinberg continued. 'On the one hand it's nice to feel confirmed in your findings. But on the other hand you begin to feel the hot breath of competition on the back of your neck!' His audience chuckled. 'But he had unalloyed pleasure at this – which I sort of registered and then tucked away.'

The Canadian scientist had obviously impressed a lot of people, because not long after this he was invited to give talks at the MIT Cancer Centre, Cold Spring Harbor and other prestigious labs in the US. 'But just before he was to come,' said Weinberg, warming to his story, 'David Baltimore swung around from his office to mine and said, "You won't believe what happened . . . this guy's boss in Toronto has just thrown him out of the lab!" I said, "*What*?" He said, "Yeah, they sent a paper to *Cell* to be reviewed, and one of the reviewers calculated how many Petri dishes would need to have been used in order to carry out this work – he calculated that it was more Petri dishes than were used in all of eastern Canada that year!"' Weinberg pulled a face to express his incredulity, and his audience laughed.

'And so it turned out – although the guy to this day never admitted any wrongdoing – that this idea he had was maybe right on the mark, but all the data he published came out of his brain rather than his lab bench . . .'

The result of this fraudulent behaviour was that, for the next 10 years or so, labs like Weinberg's had difficulty getting findings from similar experiments accepted by the cancer community, which was wary of being duped again. But as evidence for oncogenes began to accumulate from many different labs, doubts about their central role began slowly to fade. The research community became fixated on an 'accelerator' model of cancer – one in which the normal mechanism of cell division is being actively reprogrammed by these 'rogue' genes, the oncogenes, to go into overdrive, thus causing the cells to proliferate wildly. This was the mindset at the time p53 was discovered in 1979.

CHAPTER THREE

Discovery

In which we: a) meet the scientists who stumbled across p53 while investigating the cancer-causing oncogene in a monkey virus; and b) hear how, every time they tried to purify the protein made by the oncogene, they found another protein in their test tubes that they couldn't shake off.

***

How uncertain it can be, when a man is in the black cave of unknowing, groping for the contours of the rock and the slope of the floor, listening for the echo of his steps, pushing away false clues as insistent as cobwebs, to recognise that an important discovery is taking shape.

Horace Freeland Judson

The history of science is strewn with groundbreaking discoveries which are only subsequently recognised as such. No drama attaches to the moment itself, and life goes on as it was before. Often, the circumstances are mundane – a scruffy laboratory with test tubes and microscope slides scattered among scientific papers, family pictures and a postcard from a colleague on holiday pinned to the wall; a white jacket over the back of a swivel chair. The discovery of p53 in 1979 was no different. It occurred independently in labs in London, Paris, New Jersey and New York at almost exactly the same time, though the two men most widely credited with finding the super-gene are David Lane and Arnie Levine, who published their research in the most prestigious journals, *Nature* and *Cell* respectively.

Both were working in the still somewhat contentious field of oncogenes – Lane at the Imperial Cancer Research Fund (ICRF) in London and Levine at Princeton University,

New Jersey. After reading through the dry-as-dust accounts of their findings in the scientific literature, I set off to hear their stories from the men themselves. But first some information about the experimental system they were working with, which has a colourful story of its own. They were studying a virus called SV40, which stands for simian vacuolating virus 40; it is the workhorse of molecular biology because it provides a simple model for exploring how the machinery of cells – DNA, genes and proteins – works in complex organisms, including us.

As its name suggests, SV40 infects certain monkey species, though it does not usually cause them disease. The virus was discovered in 1960 by American microbiologist Maurice Hilleman, as a contaminant of polio vaccines. The original Salk and Sabin vaccines were made using kidney cells of rhesus macaque monkeys, native to the jungles, forests and dry plains of Asia, as a medium for growing polio viruses. The virus, which grew prolifically on the monkey cells, was harvested and treated so that it could no longer cause disease, but would induce an immune response that would protect the vaccinated individual against subsequent infection with harmful polio virus.

Hilleman, who worked for the Merck Pharmaceutical Company and was famed for his fiery temper and for keeping in his office a row of 'shrunken heads' (models made by one of his children to represent the employees he had fired for not coming up to his exacting standards), was a pioneer of vaccine research. In his lifetime he developed more than 40 vaccines, including those for mumps and measles. Hilleman had warned of the possibility that monkey cells used to grow the polio virus might contain other viruses and, five years after the first mass vaccination campaigns against the paralysing disease began, he and his colleague Benjamin Sweet duly showed that they did.

By the time the rogue virus harvested alongside the polio virus was detected, millions of people worldwide had received their shots. But no one seemed unduly concerned until the following year, 1961, when reports appeared that injection of SV40 into newborn hamsters caused tumours to develop. The US Government ordered all new batches of polio vaccine to be screened to eliminate SV40. Existing stocks were not recalled, however, and by the time these were exhausted in 1963, 10–30 million Americans and countless people in other parts of the world had been injected with contaminated polio vaccines (the oral vaccine was not a threat). But despite exhaustive investigation in the US and Europe over more than half a century – and the detection of traces of SV40 in some rare tumours of the brain, bone and lung – public health watchdogs such as the Centers for Disease Control and Prevention and the National Cancer Institute in the US assert that there is no evidence of increased risk of cancer in people who might have received contaminated polio vaccine. Nor, they say, is there any evidence that the monkey virus found in the rare tumours was the cause of the cancers – though suspicion that it was will probably never completely go away.

ONE MAN'S JOURNEY OF DISCOVERY

For David Lane, who began working with SV40 in the mid-1970s, this was a completely new field. Trained in immunology at University College London, he was still trying to write up his PhD thesis for his charismatic professor, Avrion Mitchison, when he was asked to join the laboratory of Lionel Crawford at the Imperial Cancer Research Fund. Besides offering the excitement of a new intellectual challenge, cancer research had emotional significance for Lane. During his first year at university, his father had died of colorectal cancer within six months of complaining of backache that would not go away. He left a still-young wife and five children ill-prepared to face such loss. 'It was very fast and very shocking,' Lane said as we sat talking over coffee in the conservatory of his home in Scotland, the late-summer sun streaming into the room and the occasional ear-splitting roar from a jet fighter streaking across the sky from nearby Leuchars air base. He cared for his father in his last weeks. 'I saw the disease as it really affects people, and that was a pretty formative experience. I felt it very strongly – we all did as a family.'

As an immunologist Lane had learnt how to use isotopes of iodine to label proteins within cells so that they could be watched in experiments, and it was this skill that Crawford's lab at ICRF was after. 'Incorporating the isotope makes proteins very radioactive and screamingly easy to detect,' he told me. 'I was good at this; I had to do it for my PhD. But not everyone likes handling these materials. I think it's a completely safe isotope, but it's kind of noisy. I mean, the Geiger counter goes *crrkk!* And you know you've done the trick!'

The facts that SV40 can 'transform' normal cells into cancer cells and that its effect is dramatic are what made it attractive to researchers at the ICRF. Just before Lane went to work there, the viral gene responsible for transformation, SV40's oncogene, was identified as something called 'large T antigen'. To begin to tease out exactly how the virus transforms cells when it infects its host, scientists needed to study the protein it produces when the large T antigen is switched on, and this is what Crawford's lab was doing. Lane's task was to develop reagents, or tools, to highlight and extract the large T antigen protein from infected cells. The challenge was to obtain manageable quantities of the stuff that was pure and separated from all the other gunge in the cells. The reagent Lane developed worked on the same principle as the immune system. Just as our bodies design antibodies to seek out and 'capture' invading foreign particles for destruction by scavenger cells, so he designed antibodies to recognise specifically large T antigen.

'It was a tremendously exciting time,' recalls Lane, a

tall, outgoing man with boyish good looks and the kind of good-humoured enthusiasm that says that, no matter what the challenges, the pressures and the politics, science for him has always been fun. 'I felt we were right at the centre of things. We were on this fantastic floor at the ICRF with the real pioneers in the field. Renato Dulbecco [who had just won the Nobel Prize for Medicine for his work with tumour viruses] was there, and I remember Harold Varmus came over on sabbatical. The atmosphere was very intellectual, very knowledgeable; lots of meetings; lots of critique of your data.'

Working together on the development of their reagents for harvesting purified large T antigen, Lane and Crawford had just begun to make progress when the older man went for a year's sabbatical in the US, leaving his newest recruit effectively in charge of his lab. Lane, still in his early twenties, was plunged into the office politics, with a number of people angling to increase their lab space and exercise their authority by telling the young scientist what to do while his boss was away. But Lane was not to be diverted. Not only is he naturally resistant to being bossed, but he had spent his formative years under the tutelage of the brilliant, unconventional and irreverent Avrion Mitchison, nephew of the eminent British biologist J B S Haldane and a man whose teaching philosophy was to let students follow their own noses (according to Lane, his professor once set an exam for his students by simply putting out a row of objects on a bench and asking them to comment – a test of their imagination that fazed those who had swotted up on the questions they had expected to be asked). The years with Mitchison had strengthened Lane's independent spirit; at the ICRF he ignored his senior colleagues' attempts to redirect his research, and just got on with what he wanted to do.

'We made what we thought was a terrifically good, specific reagent to large T antigen,' he recalled of the period leading up to the discovery of p53. 'We were really happy with it because we'd done all kinds of clever things to make it right. But when we used it, instead of – as we'd hoped – just bringing down the one protein, it always brought down this other protein, with a molecular weight of 53 kilodaltons, as well.'

The process he used to harvest the large T antigen was called electrophoresis, which involves passing an electric current through a gel sandwiched between glass plates to which the protein mixture has been added in little wells. The current causes the protein molecules to migrate through the gel according to their size and electrical charge, thus separating them; the large molecules don't migrate far, while the little ones go a long way. But no matter how hard Lane tried, he never seemed able to get the large T antigen pure and simple on its own: there was always this nagging 'shadow' in the gel.

To those of us unfamiliar with the microscopic world, it's hard to imagine getting excited by two small, dark smudges on a plate of gel; or that such a mundane image might suggest the beginning of something momentous. But like beauty, scientific discovery is in the eye of the beholder: it depends on the scientist himself or herself seeing something singular in what to other eyes may seem commonplace or dull, and then having the wit to know what it might mean. To draw an analogy from the macroscopic world, the Laetoli footprints would have been nothing but patches of displaced dust on the floor of a remote African canyon to the untrained eye; but to the palaeoanthropologist Mary Leakey, who discovered them in 1976, they were a window into the fathomless past and thrilling revelations about the origins of humankind.

Those tiny dark smudges in his laboratory gel certainly excited David Lane, even though he could not, at that stage, have known what he had found, for p53 turned out in the fullness of time to be a type of protein never before seen. Others in Lane's lab were unimpressed: it's a contaminant, they told him; his tools were not as good as he imagined, some suggested; or perhaps the 'rogue' protein was a breakdown product of the large T antigen which was being chopped up into little pieces in the infected mouse cells he was using for his experiments. But Lane trusted his tools, antibodies designed to recognise and attach themselves to large T antigen and to no other protein; he had been extremely careful to avoid contamination of his experiments; and he remained convinced that what he saw repeatedly in those gels was important. 'I was very sure in my mind that it was something to do with how the virus transformed cells, because I was kind of primed for that. I mean, Joe Sambrook (a highly respected tumour-virus specialist) had written an article saying, "Look, there can't be many ways large T antigen works. It's a single protein that goes into a cell, it transforms the cell into a cancer cell; it *must* be interacting with some part of the host machinery." So I was looking for exactly that.'

Here again Lane's background in immunology came in useful, for it told him that if, in the infected cell, the two proteins were sticking to each other physically, an antibody that recognised one would automatically bring down the other – 'a sort of piggyback idea'. It strengthened his conviction that the interaction between the two was central to the way large T antigen turned the cells cancerous. He was able to get enough evidence to support that view and to excite the imagination of Lionel Crawford, newly returned from sabbatical, who repeated his experiments. The two published their findings in *Nature*, which was sufficiently impressed to make this the cover story of the journal of 26th April 1979.

OTHERS ON THE SAME TRACK

Meanwhile, across the Atlantic, in a lab surrounded by quiet leafy parkland at Princeton University, Arnie Levine was also studying the monkey virus SV40 in an attempt to uncover the mechanism by which cells turn cancerous. 'Ever since I was a kid I've been fascinated by viruses,' Levine told me when I visited him in New Jersey, taking the train from Penn Station out through New York's scruffy post-industrial suburbs towards the prosperous, sedate university town of Princeton. 'What caught my imagination was that for a hundred years the effects of viruses were known, but no one had ever seen them. They're the smallest of all living organisms, and they're completely degenerate!'*

Levine, a genial man now in his sixties, grew up in New York City. His father owned movie theatres, and from time to time the young Levine would earn pocket money working as an usher and clipping tickets. 'I must have seen James Cagney in *Yankee Doodle Dandy* about 19 times,' he chuckles. 'I grew up in a wonderful neighbourhood in Brooklyn, a typical American neighbourhood with Italians, Scandinavians, Jewish people – just a real mixture. My grandparents came from Poland and Lithuania. My father actually came from Lithuania as a young boy, but he never went to school; he went right to work, as was the case with most immigrants because they needed to earn money to survive.'

His parents believed fervently in educating their children, and the young Levine went to Harpur College (now Binghamton University) in New York City, where he was taught microbiology by Mildred Shellig, a retired medical doctor whom he credits with inspiring in him his passion for science. 'Dr Shellig's enthusiasm was infectious, and what I specially didn't expect was that I *loved* the laboratory. We got to repeat people's experiments from the literature and even try some new things. People stayed in the lab late

---

* 'A virus is nothing but a package of genes inside some proteins. So whether it's alive or not is kind of debatable. It's either a kind of a complex chemical or a very simple life form,' says Jeffery Taubenberger, Senior Investigator in the Laboratory of Infectious Diseases at the National Institute for Allergy and Infectious Diseases, Bethesda, Maryland.

at night; she had us over to her house to talk, and what did we talk about? We talked about science! It was terrifically infectious for me. This was probably 1959 – only six years after Watson and Crick and their discovery of DNA structure. So the molecular biology revolution was just starting.'

Levine began his career studying how viruses replicate themselves – essentially by taking over the machinery of the host cell to do so, because they are parasites that cannot function outside of other living things, be they plant or animal. He was working with bacteriophages – viruses that infect bacteria, described earlier. They were exciting model systems because everything happens so fast: an infected bacterium can spew out a new generation of virus roughly every 20–60 minutes, compared with an animal cell where the process might take 48 hours. But as the story of oncogenes and animal tumour viruses took off, Levine switched his focus. His imagination was fired. 'How is that possible?' he wondered of these minuscule scraps of life. 'How can one or two simple genetic elements which encode the information for proteins cause a cancer? It seemed to me to be the way into understanding, for the first time, the origins of cancer in humans. So my movement into this field was really based on the use of the simplest system to get at one of the most complex of questions: what is the origin of cancer in us?'

One of the theories around at the time Levine discovered p53 was that cancer cells become reprogrammed to resemble embryonic or fetal cells. In other words, the evolutionary clock is turned back to a stage where the cells' natural behaviour is to grow and divide rapidly, as they would in a developing embryo. Researchers had found evidence for this in the presence of regressive fetal proteins in liver and colon cancers, and had developed blood tests to detect the antibodies that are created by the adult immune system recognising them as aberrations – out of time and out of place – and arming itself to attack them.

The 're-embryonisation of cancer cells' was an attractive

concept because of the obvious behavioural similarities between the two cell types, embryonic and cancer, and the hunt was on in a number of labs to identify proteins that were present in both normal embryo cells and tumour cells, but *not* in healthy, fully developed adult cells. These proteins, the researchers figured, while obviously good news for developing embryos, could be the driving force behind cancer. In his lab, Levine's team was looking for evidence of a fetal protein or proteins expressed in response to infection with SV40, and when p53 appeared Levine believed at first that was exactly what they had found. What specially excited him when his graduate student Daniel Linzer, who did the original experiments, showed him his results was that the rogue protein occurred in large quantities in the SV40-infected cells, suggesting it must be doing something important, and that it was interacting specifically with the viral oncogene, large T antigen. What's more, his team had found exactly the same protein also in uninfected fetal cells.

By the time they published their paper in *Cell,* Levine and Linzer knew from further experimentation that p53 was not the fetal protein they had been looking for, and over time the theory itself fell from favour. But apart from the fact that it was obviously important because of its regular appearance in cancer cells, Levine did not know what to make of the new discovery. 'We had no idea at the time where this would go compared to where it went – that's for sure!' he chuckles, pausing to reflect. 'I don't think any of us thought this was going to be the single most important gene in cancer based on the frequency with which it mutated . . . However, I *would* say we thought we had found the path into how SV40 causes tumours.'

THE PARIS GROUP

In Paris, a third group of scientists who also independently discovered p53 were equally mystified by the protein they found sticking close to large T antigen in their experiments with SV40. The discovery occurred in the lab of Pierre and Evelyne May at the Integrated Cancer Research Institute in Villejuif, where researchers were working simultaneously with SV40 and another closely related virus called polyoma, which causes multiple tumours in animals. Besides having a large T antigen like SV40, polyoma also has a small T and a middle T antigen; these are involved in transforming the cells the virus infects, with middle T antigen being especially powerful at causing tumours. Most people working in the field assumed that SV40 also had a middle T antigen that was likely to be equally potent; this was in the minds of Pierre May's team when his doctoral student, Michel Kress, found large quantities of a new protein in his experiment.

Thierry Soussi, who was to become an important figure in the unfolding story of p53, was working in another lab along the corridor from the Mays in 1979, studying SV40 replication. 'I vividly remember a postdoc* bursting into our laboratory to announce that his friend, Michel Kress, had identified the middle T antigen of the SV40 virus: it was a 53 kilodalton protein,' he wrote in a brief review of p53's history for a molecular-biology journal in 2010. But when further investigation revealed that the new protein Kress had discovered in his SV40-infected cells came not from the virus but from the host, no one knew quite what to make of this. He and the Mays published the findings without embellishment in the *Journal of Virology*, read by few people outside this specialist field.

Pierre May died in 2009, and Evelyne May and Michel Kress are retired, so when I visited Thierry Soussi at the Karolinska Institute in Stockholm where he now works, I asked about the discovery of p53 in Paris. Ushering me into

* A postdoctoral scholar ('postdoc') is an individual with a doctoral degree who's engaged in a temporary period of mentored research and/or scholarly training in order to acquire the professional skills needed for his or her future career.

a bright, modern office opposite his lab, Soussi turned off the opera he likes to listen to while he works, pulled an old file from a high shelf and opened it, blowing away the dust as the scent of musty old paper wafted from the typewritten pages, yellowed with age. This was a copy of Kress's original doctoral thesis and his paper about p53. Soussi leafed through it with a faint air of regret: 'Michel Kress has been forgotten, which I think is a pity. He has been forgotten for two reasons. First, he is not ambitious at *all*. Second, he discovered p53 and one year later he had to go for a postdoc, and he went to a very good lab. He wanted to work on p53, and this lab told him "p53 has no future; you are going to work on something else." Therefore he had to give up on p53 right at the beginning – which was not too much of a problem because at this time no one was really believing anything about p53. No one could be excited by something when you don't know what it is.'

Except, that is, for Pierre May who, according to his widow speaking to me on the phone from Paris, had a hunch from the beginning that this would turn out to be significant. Over the following decades May won several prestigious prizes for his work on the gene, though at the time of its discovery in 1979 he had the greatest difficulty raising funds to continue the investigation.

Ironically, the lacklustre response to p53 is exactly why Soussi himself chose to study the gene when he joined the Mays' lab in 1983: he figured it was just the kind of quiet, uncompetitive backwater that he could cope with alongside the heavy burden of teaching he was expected to fulfil as a university researcher. He smiles at the thought of how wrong he was – but he wasn't alone. 'You know, I was working with a student at this time who was doing her thesis on p53 and she wanted to apply afterwards to INSERM (Institut National de la Santé et de la Recherche Médicale), which is the French equivalent of the National Institutes of Health. On the board for her thesis we put the director of INSERM, and at the end he told her, "Okay, I understand that you want a position here. There should be no problem: you have a good application. But just don't work on this bullshit protein; change your topic." This was exactly his word! No one honestly could anticipate in the early 1980s what p53 would become – it was impossible.'

In spite of his own personal conviction that p53 was important – perhaps even key to the development of cancer – David Lane faced the same kind of prejudice in the early days. Soon after his discovery, he spent a few months at Cold Spring Harbor laboratory on Long Island, New York, where James Watson – just one of many Nobel-winning scientists who had worked there – was director. Unimpressed by Lane's recent discovery, one of his new colleagues predicted he would one day be ashamed of the claims he had made for p53's significance.

A DIFFERENT ROUTE TO THE SAME DISCOVERY

The fourth person to discover p53 in 1979 was Lloyd Old, who died of prostate cancer in 2011. Born in San Francisco, California, in 1933, Old started his working life as a professional violinist, studying the instrument in Paris and at the University of California, Berkeley, before his fascination with science overtook his musical ambitions and he switched to medicine. Old was a pioneer of tumour immunology, which studies the interaction between our bodies' immune system and cancer cells. It was while he was trying to identify what it is about certain tumour cells, but not others, that alerts the immune system and causes it to develop antibodies tailored specifically to recognise only those cells – one of the central conundrums in tumour immunology and infernally hard to crack with the tools available at the time – that he discovered p53.

As with the virus studies, the rogue protein seemed to piggyback on other proteins that Old and his team at

Memorial Sloan Kettering in New York were trying to isolate by using specially designed antibodies as tools, rather as you would use a magnet to pick out scraps of metal from a bunch of other materials. This was intriguing and, working with laboratory mice, they looked for the protein in all kinds of cell types, both normal and cancerous. The researchers found p53 in none of the normal cells, but in all of the cancerous ones, and concluded it must be playing a part in the cancer process.

Old and his colleagues published their findings in the *Proceedings of the National Academy of Sciences* (*PNAS*), but because there was – and still is to a lamentable degree – so little communication between different fields of cancer research, it was over a year before the immunologists and virologists realised they were all talking about the same thing. But what exactly was it that had caught the attention of such a disparate bunch of scientists, all following different leads in their widely scattered labs, at about the same time? This was the burning question now for the cancer community as they began to realise that p53 could not be dismissed as a contaminant or an irrelevance, but may well be key to the transformation of cells. To find out, they needed to clone the gene.

CHAPTER FOUR

Unseeable Biology

In which we peer into the machinery of the cells to see how the genes make the proteins that do virtually all the work in our bodies.

***

Every cell in nature is a thing of wonder. Even the simplest is far beyond the limits of human ingenuity. To build the most basic yeast cell, for example, you would have to miniaturise about the same number of components as are found in a Boeing 777 jetliner and fit them into a sphere just 5 microns across; and then you would have to persuade that sphere to reproduce.

Bill Bryson

The early 1980s were an incredibly exciting time in biological research, with the increasing ability to clone and sequence genes providing a tool of huge importance. Here a bit of basic biology is needed to make the next step in the p53 story intelligible. Virtually all the activity in our bodies is performed by proteins, and these are produced, or 'encoded', by genes, which are, in effect, recipes for all the different proteins. The proteins are made only when and where they're needed, at which time the relevant gene is switched on. And there are mechanisms for removing the proteins when they have finished their tasks. When its protein is not needed, the gene sits there in the cells, quietly doing nothing.

Sequencing a gene gives, as it were, the exact recipe for the protein it encodes, and provides vital clues to its purpose and function in the cell. Moreover, having the clone of a gene – that is, endless copies of the same thing – to work with means that all sorts of experiments can be done in cultured cells in the lab to answer questions about how the gene might work. Today, cloning is a relatively simple process that can be accomplished in a day or two, but in the early 1980s it was a big challenge taking months – even years – and was made especially difficult by the fact that it involved working with recombinant DNA. This technology remained somewhat controversial for years after the Asilomar Conference, which, you will remember, brought concerned scientists together in a California hide-out to confront the spectre of Frankenstein species.

It's worth pausing at this point to ponder what it is that molecular biologists – working away with their pipettes and dishes, test tubes, gels and incubators – are handling. What can they actually see? What can they feel and smell? And what faculties are they using beyond the basic senses to understand what's going on? 'The wonderful thing about molecular biology is that it's based upon faith, ultimately,' says Peter Hall. 'All the data fit with a model; you build it and you test it, but you can't *see* it. Oh no. No, no. You actually infer it from all the experiments that you do, the interpretation of which is this, that or the other.'

It's the 'unseeable' nature of molecular biology – because most molecules are smaller than the wavelength of light – that makes it so difficult to grasp and therefore so intimidating to the layperson. But if, like a deep-sea diver, you're prepared to learn how to operate in an alien environment, there's a fabulous world to be explored. So, in the best tradition of scientists themselves, let's translate what has been deduced from experimentation about the basic working of cells into mental images and concepts that will make it more comprehensible.

First of all, it's not strictly true that everything in molecular biology is invisible. With an electron microscope, under certain conditions, you can actually see DNA, which is a 'mega-molecule'. It appears sometimes as a fine strand in cells, like a piece of cotton thread. But it is when the thread is compacted, spooled tightly on to structures called histones and organised into paired chromosomes as a cell prepares to divide that it is most easy to see, if a stain is used to highlight it in the cell. However, no microscope currently available to most labs can show DNA in enough detail for scientists to be able to determine the order of the 'bases' making up the molecule. Thus the genes which are carried on the chromosomes – not as discrete chunks of DNA but as segments along the continuous stand of genetic material – remain unseeable, and it is these scraps of information that carry the recipes for the proteins that the scientists are after.

The famous corkscrew structure of DNA – the double helix discovered by James Watson and Francis Crick in 1953 – is made up of components called nucleotides, which stack one on top of the other like nano-sized blocks of Lego to form long chains. Each nucleotide, or block, has three components: a sugar molecule called a deoxyribose (the D in DNA), a phosphate group and a nitrogen 'base'. These bases come in four different types: adenine (represented by A), thymine (T), guanine (G) and cytosine (C). The long spiral ribbons of DNA are double-stranded, and the bases on one strand reach across to pair up with bases on the other, holding the two strands together like the rungs of a chain ladder. No matter what the organism, the bases of their DNA always pair up in the same way: A with T; G with C.

DNA is measured in base pairs. Human DNA is around 3 billion base pairs long and we have about 1.8 metres (6ft) of it in every one of the trillions of cells in our bodies, apart from the mature red blood cells, which are unique. You get an idea of the scale of the landscape in which molecular biologists are working when you consider that, if all the DNA in your body was uncoiled and laid end to end, it would stretch to the moon and back more than 3,000 times. This unimaginably long gossamer thread is the instruction manual for building and operating the machinery that is you or me, so scientists working on the Human Genome

Project, which aimed to decipher the manual's code, were staggered, and not a little mystified, to find that the genes, the working units of DNA, account for only around 2–3 per cent of the stuff – that is, 24,000 genes, each averaging 10,000–15,000 base pairs in length. Having no clue to the purpose of the rest, they labelled it initially 'junk DNA'.

That was in 2000. Since then researchers have worked out the function of approximately 80 per cent of what is now more respectfully called 'non-coding DNA'. That function is to regulate the expression of the genes – when and where they are turned on, and at what volume. Since every cell has an identical complement of DNA and a complete set of genes, these switches are vital to the way cells can be so different – a liver cell from a heart cell, a brain cell from a skin or bone cell, for example. The 'switches' of the non-coding DNA ensure that only the genes relevant to each organ are activated in the cells of that organ and the other genes left silent.

When a gene is switched on, its recipe is read out to produce a protein. For this to happen, the two strands of DNA need to be separated so that the genetic information, which is on the inside, is exposed. Here I'm going to switch the analogy for DNA from a spiral chain ladder with rungs to a zip-fastener with teeth. A little machine called a helicase unwinds the DNA on the chromosome carrying the relevant gene, and then unzips it so that the recipe can be read. The DNA is housed in the nucleus of the cell, but the proteins are made in the body of the cell, the cytoplasm outside the nucleus. Because the DNA molecule is too big to pass through the pores of the nuclear membrane into the cytoplasm, the genetic information is copied, by means of another little machine, an enzyme called polymerase, on to a smaller molecule called messenger RNA (mRNA). This forms a single complementary strand to the DNA. The mRNA leaves the nucleus and heads for the protein assembly factory, the ribosome, in the body of the cell.

Proteins are made of amino acids, of which there are 20 different kinds. The instructions from the gene determine which amino acids are to be used, in what order and what quantities. For this process, the string of simple letters represented by the sequence of base pairs on the gene, the As, Ts, Gs and Cs, need to be translated into more complex 'words', so the ribosome protein factory reads the linear information on the strand of mRNA in three-letter chunks. These chunks, or 'words', are called codons, and they identify which amino acids should be used to make a specific protein. The amino acids are brought to the assembly point, the ribosome, from other parts of the cell by transfer RNAs (tRNA), which dock on to the appropriate codons and deposit their cargoes of amino acids. When all the amino acids are in place they are strung together as a chain, which detaches itself from the strand of mRNA and goes off to another part of the cell to be folded. This process is vitally important, since a protein's function is determined not just by its amino acid components but also by the way it is folded.

The other biological process central to the story of p53 is that of DNA replication, which occurs in every cell that is about to divide. In this process once again the enzyme helicase unwinds the DNA and unzips it – not all the way along, but a fragment at a time. Small molecules called single-strand binding proteins, or SSBs, attach temporarily to each separated strand to stabilise them while they are being copied and to ensure that they stay separate. Then DNA polymerase travels along each separated strand attaching new nucleotides – the nano-scale Lego blocks described earlier – to each of the existing ones, pairing up the bases in the conventional way, A with T and G with C, thus constructing a parallel strand of DNA, block by block. A subunit of the polymerase travels along behind, 'proofreading' the new DNA to see that it has been faithfully copied. Then an enzyme 'glue' called DNA ligase seals the fragments of copied DNA into a continuous double-sided strand that rewinds itself automatically.

In the replicated DNA, one strand of the double helix will be from the original (known as the parent strand) and the other will be the new copy (known as the daughter strand). The cell is now ready to divide into two cells with equal shares of identical genetic material. This process, going on ceaselessly in billions of cells in our bodies as we repair and replace tissue and our hair and nails grow, is so efficient that mutations – mistakes that escape the proofreader – occur at the rate of about one in 10^9 nucleotides per replication.

It's interesting to note that this knowledge, this understanding of how the machinery of life works, is built on the foundations of Watson and Crick's discovery of the structure of DNA. The double helix – the spiral staircase drawn originally for their paper in *Nature* by Crick's wife Odile – remains one of the iconic images of 20th-century science. Yet when the paper announcing their discovery came out in 1953, it was noticed by hardly anyone beyond a small group of enormously brainy and ambitious scientists working in the same field – some of whom had been racing to make the discovery themselves first.

Of the media, only one paper, the British *News Chronicle*, carried the story about 'an exciting discovery about what makes YOU the sort of person you are . . .' And for nearly a decade, only a tiny proportion of scientists writing about DNA in professional journals mentioned the double helix. It was a beautifully elegant model, but many biochemists, intensely preoccupied with working out how we synthesise the proteins in our cells, were sceptical that genes – still rather an abstract notion in the early 1950s – had anything to do with it.

CHAPTER FIVE

Cloning the Gene

In which we hear about the huge technical challenge and the hot competition to clone p53 as the first step to discovering how the gene and its protein work.

***

We stand on the wrong side of the tapestry – a confusion of colours, knots and loose ends. But, be assured, on the other side there is a pattern.

Anon

The cool reception and slow build-up of recognition for the double helix – culminating in the Nobel Prize for James Watson, Francis Crick and the biophysicist Maurice Wilkins in 1962 – are instructive. This is how science works, says Peter Hall. 'It's the analogy of the man in the dark warehouse with a small pen-torch. He can only see a tiny part of what's there. The whole thing only becomes clear when he gets more light.'

Until they had a clone of p53, the scientists were guddling around in the dark without even a pen-torch, and in the early 1980s the race was on among a handful of individuals in labs around the world, from the UK and US to Russia and Israel, to clone, or make identical copies of, the gene – the first, essential step to finding out what it is and how it works. The first person to succeed was Moshe Oren, who started the process while working as a postdoc in Arnie Levine's lab in Princeton, but completed it at the Weizmann Institute in Israel, where I went to talk to him on a hot October afternoon some three decades later. Entering his small office on the top floor of the institute, with its views up into the wide sky, I was greeted by the scent of citrus. Oren was seated behind his desk with a little pile of orange peel and pips in front of him – clementines picked from his own garden that morning, he told me, as he offered me a handful. We sat eating the sweet fruit as we talked.

'I was looking for a new project because I needed to change. Here was this interesting protein that people were beginning to look at and it was my chance to clone something,' he recalled. 'In those days cloning was a major technical challenge. It was probably two and a half years from me saying, okay, I'm going to clone p53 to actually getting the thing. It took a lot of setting up protocols and testing different approaches, some of which didn't work. It's kind of amazing how we've progressed: cloning a gene now is trivial; it's become probably a high-school exercise! But when we did it the tools were very limited; there were very few genes that had been cloned, and each of them was cloned by a variety of improvised tricks . . . It was not easy at all.'

One of the hardest tasks for the would-be cloner even today is identifying the individual genes on a continuous strand of DNA – where does one gene end and the next begin? Oren's strategy was to look for the gene after it had been switched on, when the relevant segment of DNA had been copied, or 'transcribed', into 'messenger' RNA (mRNA), left the nucleus and gone to the ribosome – the protein-making factory – in the body of the cell. Using antibodies tailored to recognise the p53 protein from among a mass of proteins being made at the same time, he isolated the relevant protein factory and scraps of mRNA. Using the mRNA as a template, he synthesised 'complementary' DNA (cDNA) by attaching new nucleotides, block by block, along its length, pairing up the bases as appropriate. This cDNA, he hoped, would give a faithful copy of the p53 gene – or at least that part of the gene responsible for making the protein.

In order to multiply the little scraps of cDNA, he transferred them to a bacterium – E. coli, which, you will remember from the Asilomar story, is one of the workhorses of biotechnology because it's easy to manipulate, efficient at taking up new genetic material and can pump out clones at a terrific rate, given a good food supply. In order to transfer the cDNA to the bacteria, Oren had to use a suitable vector – something that would breach the walls of the bacteria and take up residence inside without killing its host – and for this he chose a plasmid.

Plasmids are tiny rings of DNA between 1,000 and 10,000 base pairs long that some bacteria have floating around in their cells, independent of their regular genomes. When a bacterium dies and its body – a single cell – disintegrates, the plasmids are scattered into the environment and are often absorbed into other bacteria, which then begin to express the new traits encoded by the plasmids. This is how recombination – the phenomenon debated at Asilomar that still excites such public controversy over biotechnology today – happens in nature, and it has provided a marvellous tool for cloners. The scientists put the plasmids in solution with the scraps of cDNA they want copied, having used a 'snipping' tool, an enzyme, to cut a gap in the plasmid ring where the new material should be inserted. With a little coaxing, the cDNA moves of its own accord into the gap, where it is glued into place by 'repair enzymes' added to the solution. This plasmid is now a recombinant DNA molecule – a mixture of genetic material from different organisms. The cloner then adds it to another solution containing the E. coli, or whatever else he or she has decided to use as a 'clone factory', and waits for it to find its way in among the machinery.

'On paper it looks very simple,' said Oren, shaking his head slowly at the memory of the months of trial and error and frustration working mostly in the realms of unseeable biology. 'In practice, we spent a year trying things that didn't work. Then we tried this method that worked, but we went through many false negative clones before we found one that was real.' The trickiest part was getting the plasmid vectors to take up their new cargo, the scraps of p53 cDNA. Oren and his technical assistants screened the plasmids 10 at a time rather than individually to see if they had succeeded – a task of enormous tedium that swallowed nearly 18 months of their lives.

'The procedure ended up being rather inefficient and the percentage of plasmid clones that indeed contained p53 cDNA was extremely low. We had to screen many hundreds of clones – actually I think about 1,800 – before we hit the first positive one.' When that finally happened, it was one of the most exciting moments in Oren's scientific career, he told me with a smile. He and his colleagues published their results in *Proceedings of the National Academy of Sciences* in 1983.

Meanwhile, back in Princeton, Arnie Levine had started collaborating with the pharmaceutical company Genentech and one of their expert cloners, Diane Pennica, to continue the search for a p53 clone once Oren had returned to Israel. Using different strategies and different cell types from Oren, they too were successful and they published their findings in the journal *Virology* the following year.

Others were hot on the same trail and by 1984 there were a number of p53 clones, of different cellular origin, in the scientific press and in limited circulation around the labs. Among them was one obtained at the ICRF in London by John Jenkins, one of very few people with cloning skills in Britain, where progress in manipulating genetic material had been inhibited by the controversy surrounding recombinant DNA in Europe and the US. Jenkins, who came late to science after finding himself a misfit at school, leaving early and taking up work as a general labourer and landscape gardener before returning to education, remembers fierce competition to obtain a clone. 'Obviously people are driven by their career ambitions – but at some level there's the pure joy of beating the opposition. I mean, it's a competitive business! Always was; always will be.' Like Oren, Jenkins remembers how tough a task cloning p53 was. This was the cutting edge of science: 'Everything at the cutting edge is a challenge at the time,' he commented.

Just how much of a challenge cloning was in the early 1980s is hard to grasp for succeeding generations of biologists. 'My best story about that was years later,' said Jenkins. 'One of the people in my lab was helping me clear out my office. We came across a bound copy of my PhD thesis and she said, "Oh! Can I take this away and read it?" I said, "Sure, no problem. Bit of old history . . ." Later she brought it back and said, "I don't believe it! You didn't do anything, and you got a PhD for *that*?" I said, "Listen! You try doing that work when you have to make all your own enzymes . . . There weren't any kits in those days, you know . . . Everything you take for granted now was just not there, so it took a long, long time – and it was tricky."'

Oren, Levine and Jenkins had, in fact, been beaten to the finishing tape in the highly competitive race to clone p53 by a Russian scientist, Peter Chumakov, working at the Institute of Molecular Biology in Moscow. But Chumakov had published his results – in December 1982 – in *Proceedings of the Academy of Sciences of the USSR*, a Russian-language journal with limited readership in the West, and for some time his triumph went unnoticed beyond the Soviet Union. Back in 1979 Chumakov had just completed his doctoral thesis on SV40 and had been looking for a project that would reveal how large T antigen caused cancer when he heard of the discovery of p53. Excited by the idea that p53 might be the key to the transformation process, he decided to clone the gene to provide material for further studies.

But Chumakov faced even greater challenges than his Western competitors. His lab in Moscow had shortages of some of the most basic equipment, such as test tubes and pipettes, which he cleaned and re-used over and over again. He also had to send off to a scientist in Texas, Elizabeth Gurney, for a sample of the special antibody to p53 she had made and which he realised was vital to his cloning endeavour. He made his request without much hope of receiving the sophisticated tool. 'I thought that even if Elizabeth decided to send the sample, a package containing a suspicious test tube would be stopped either on the USSR border, or during postal censorship that occurred regularly with mail coming from the West,' he told me. But he was lucky. Visiting a colleague in a neighbouring lab one day, he noticed a small, foreign-looking package lying on a desk and he turned it over in casual curiosity. To his surprise and delight he found it was addressed to himself and contained a small phial of the antibody together with a letter from Gurney wishing him success.

His unexpected good fortune boosted Chumakov's morale enormously and he was determined to complete his task – especially as he now realised, from his new contacts with the outside world, that he was not the only one trying to clone this important gene. That he was the first to succeed was finally recognised when his paper came out in English and he began to be invited to international p53 meetings. 'There was a wonderful atmosphere of enthusiasm and hope in the lab, and we really felt lucky about being attached to those exciting discoveries in life sciences,' he commented.

For all the research teams involved, the next step after producing a clone of p53 was to determine the sequence of the gene – the words spelled out by its base pairs, A and T, G and C, that dictate the amino acid composition of its protein. 'Sequencing was not very difficult, even in those days,' commented Oren. 'It was much easier to do than cloning. But it was all manual: you had to prepare your DNA and all kinds of reagents and run gels and do everything by hand. And you had to read it and interpret it yourself. There were many errors; the method was not precise enough, but it was not a technical challenge. And by the time I got to it there was already a technician who was doing it routinely.'

With the information provided by sequencing and a

potentially endless supply of clones – some of the whole gene, others of important segments of the gene – researchers were now ready to start investigating how p53 functions within cells.

CHAPTER SIX

A Case of Mistaken Identity

In which we: a) discover that we humans also have would-be oncogenes in all our cells; b) learn that what turns oncogenes nasty is mutation; and c) hear that an oncogene is what almost everyone – erroneously – concluded p53 to be.

***

Scientists, like anyone confronting a new problem, will start with what they already know. Neither in theoretical speculation nor at the bench do they often sail beyond sight of the shore.

Horace Freeland Judson

As they set about their investigations into how p53 functions, thoughts of oncogenes were uppermost in many researchers' minds, since the first such gene in human DNA had been discovered just months earlier, in 1982. Once again Bob Weinberg, one of the pioneers of oncogene research, was there at the cutting edge, and here we need to retrace our steps a little to see how the oncogene story developed.

Weinberg was born in Pittsburgh, Pennsylvania, and raised in the United States by a mother, father and grandparents who had arrived as refugees from Europe in the late 1930s. His father, a dentist in Germany, had seen the writing on the wall as the Nazi threat grew and had been smuggling money out of the country to a brother in the Netherlands. When he finally fled to the US in 1938, he had enough money saved to retrain and re-certify as a dentist in his new home. Though the younger Weinberg had no direct experience of Fascism, the suffering of his family in Europe, where many of them did not escape or survive the Holocaust, percolated into his consciousness and he grew up with a keen sense of the precariousness of life. 'One thing my father said to me was, "It's okay to be successful and productive, but don't be too visible lest the guys in the brown shirts come in one night and take you away.'"

Although this man with the bushy moustache, self-deprecating humour and Woody Allen-ish face insists he's happiest in mud-caked overalls, pottering in his garden or splitting logs, he is undoubtedly the most visible of a bunch of scientists in three different labs who all made the momentous discovery of the first human oncogene at the same time.* Undeterred by the scandal caused by the fraudulent Canadian research described in Chapter 2, he and his team continued to investigate the phenomenon they had discovered and reported in 1979 – that rodent cells contain would-be oncogenes that can be activated by chemical carcinogens as well as by infection with viruses. By 1981 his lab, with that of Geoffrey Cooper at Harvard Medical School in hot competition, had shown that oncogenes are to be found in human tumour cells too.

Both labs had used the same ingenious and relatively new technique called 'transfection' in their research. This involves taking pure DNA from tumour cells, chopping it up and putting it in solution with normal cells that are the researcher's target. By tinkering with the solution, the scientist can induce the cells to take up some of the pure DNA floating around and to integrate it into their own genomes. He or she then cultures the cells in Petri dishes and looks for evidence that they have turned cancerous. The cells in our bodies talk to each other constantly: normal cells obey instructions that ensure orderly growth and usually stop them dividing once they have formed a single layer in the Petri dish and filled the available surface space. Cancerous cells, however, are unruly: they talk anarchy, pile on top of one another and form haphazard clumps, known as foci, in

* The others were teams led by Mariano Barbacid at the National Cancer Institute, Bethesda, and Michael Wigler at Cold Spring Harbor.

the dishes. Thus Weinberg's and Cooper's teams looked for foci – and found them.

Homing in on the specific gene or genes responsible was an infinitely painstaking process of elimination. It meant repeating the experiment over and over again with different chunks of naked DNA from the tumour cells. But the three teams racing against each other got there within a year, publishing their results in the summer of 1982. They discovered that the oncogene in the human tumour DNA was one named 'Ras'. Mirroring the story of the oncogene Src discovered by Varmus and Bishop, Ras turned out to be homologous (that is, very similar, and suggestive of a common evolutionary ancestor) to an oncogene found in a virus – in this case the rat sarcoma virus that had been discovered in the 1960s.

The following year, 1983, Mike Waterfield at the ICRF in London found that a virus that causes sarcomas in monkeys had, some time way back in evolutionary history, hijacked a gene from us that is involved in growth and repair, and is especially important in healing wounds. This was further proof that we humans have in us genes that are part and parcel of our normal DNA – and have a regular job to do in our cells – that can, under certain circumstances, cause us harm.

But what are the circumstances? At the time the first human oncogene was found, no one yet knew that carcinogens – whether they be chemicals, infections or radiation – typically work by scrambling our DNA. Bob Weinberg, however, had a hunch that mutation would be the key to activating would-be oncogenes, and once again he faced down the doubters who pointed out that some chemical carcinogens don't actually cause mutations. 'I was not troubled,' he commented wryly in an essay on the discovery of human oncogenes. 'I thought that a good simple idea should not be undermined by complicated facts.'

Weinberg's hunch proved correct, and he and his co-discoverers soon found that the Ras gene – which has a central role in orchestrating the growth, division and differentiation of our cells in the normal course of events – can be turned nasty by an alteration in a single nucleotide, representing just one letter in its protein recipe. In a sensitive position in the gene, such a mutation can result in Ras being permanently 'switched on', heedless of any signals it may be receiving, thereby driving the cell to grow and divide unchecked. It was an important revelation, another vital insight into the mechanics of cancer at the most basic level. *Nature* hailed the discovery of the human Ras gene and its activating mechanism by naming 1982 'The Year of the Oncogene'.

No one could know then just how important a find Ras would turn out to be. We now know that this gene is mutated in around a quarter of all human tumours, including roughly a half of all colon cancers and 90 per cent of pancreatic cancers. But as scientists continued to investigate oncogenes – of both viral and cellular origin – they became aware of a vital caveat: a single oncogene acting alone cannot create a tumour; in order to mess up a cell's machinery enough to cause cancer, oncogenes need to co-operate with one another. Researchers had no idea initially why this should be, only that it was so. When, for example, Ras is put into cells together with Myc – another powerful oncogene found in the DNA of a virus as well as in animals, including us – the effect is clear cut and often dramatic.

p53 *LOOKS* LIKE AN ONCOGENE

In 1984, when researchers began working with the new p53 clones to find out how the gene functions, they quickly concluded it too was an oncogene. This was one of the first ideas they tested and they did so by taking the classic experiment – pairing the powerful Myc with another oncogene – but replacing the Myc with p53 to see if it had the same effect. 'We put p53 with Ras; some others put it with other oncogenes; and we all got positive results,' said Moshe Oren. 'From the beginning the results were similar to Myc, though less dramatic and less efficient. So the feeling was: p53 is an oncogene. Not as good as Myc, but okay, Myc is the king; this is just a regular knight!'

An essential condition for cancer to get going is that cells become 'immortalised' – that is, able to override the limits set by nature to the number of times they can divide. An individual oncogene acting alone can accomplish this vital step, which leaves the cell on the brink of cancer, vulnerable to a second hit by another oncogene that will take it all the way. Jenkins' group at the Marie Curie Institute, tucked away in the southern English countryside in Oxted, Surrey, found that their p53 clone was able to immortalise cells.

Another indication that what they were dealing with was an oncogene came from the fact that it produced a dramatic abundance of protein – what the scientists call 'over-expression' – in the cells it transformed into cancer. 'At that time it was quite standard to conclude that proteins that are over-expressed in cancer cells must be produced by oncogenes,' says Oren. 'That's how viruses transform; that was in the textbook; that's what everybody thought. Varmus and Bishop got a Nobel Prize around that time for their work on Src, and the general dogma was: oncogenes are things that are over-expressed in cancer cells and that's what *makes* them cancer cells. So here we have a perfect candidate: here's a protein that's over-expressed in cancer cells, right? p53 looks *just* like any plain oncogene.'

However, while everybody else seemed to be getting nice clear-cut results with their clones, Arnie Levine back in Princeton was drawing a blank, time after time, and growing mighty frustrated. 'We saw Moshe's group publish with Weinberg on the fact that it transformed cells; we saw Jenkins publish that it transformed cells, and Crawford also; and we couldn't reproduce those results with our clone. We were devastated!' he told me. 'I had two people in the lab, a graduate student named Phil Hinds and a postdoc named Cathy Finlay, who were working on this, and to watch other people publishing and *not* be able to reproduce their results . . .'

Like everyone else, Levine's team were conducting experiments designed to confirm their hunch – based on all the bits of evidence thus far – that p53 was an oncogene, and they naturally interpreted their singular lack of results as failure, pure and simple: they had a dud clone. Something gnawed at the back of their minds, however. As the various cloners published their gene sequences, Levine's team noticed that their own clone had a difference in one codon – representing just one building block in its protein – but they had dismissed it as insignificant. Sequencing at that time was still a 'hands-on' process, where people not machines produced and interpreted results, and you had to allow a margin for human error.

'At that time a few sequencing errors were not something to be proud of, but it wasn't a major crime,' Oren concurred. 'You see a difference here, a difference there, it doesn't really matter – because by and large we can see what the protein looks like, and biologically we see that this one and that one do the same thing, so we're fine. So we didn't spend much time thinking about the significance of a difference in one amino acid. I tended to think their amino acid was wrong and mine was right, but we're dealing with the same thing, so who really cares?'

Levine and his team cared very much, as they agonised over their failure to get results. They asked Oren, as a one-time member of their lab, if they could have a copy of his clone to try their luck. Oren happily obliged, and sure enough the Princeton team reproduced his results. Now they all knew that single codon difference was significant. It began to sow niggling doubts in their minds. At one of their periodic meetings, Oren was in Levine's office in Princeton when Phil Hinds produced a list he had compiled of the sequences of all the clones the various labs had published in the literature. The only clone that had no biological effects, recalled Oren, was the one that Arnie Levine's lab had produced. 'This was very revealing because at the end of this session it was clear that all the people who were getting biological effects were using clones of *mutant* p53 – it was kind of obvious from the sequences. Arnie's was the only one that was not mutant. He had a clone of the normal (or 'wild-type') gene.

'But it wasn't that we all sat together and suddenly realised – it was something that had been cooking slowly in our minds and that finally converged in that single two- or three-hour meeting. If you compare one to one, you don't know what's right and what's wrong. You need enough information for all the bits of the Lego to fall into place and to get the feeling of what is the rule and what's the exception.'

This was a momentous revelation, but it took a while to sink in. Except among those who were still feeling the intellectual buzz of cloning and testing their creations, interest in p53 had waned as those early experiments showed it wasn't the novelty everyone had hoped for; it was 'just another oncogene', and rather a feeble one at that. This was a time when young scientists looking for jobs and promising paths ahead in the highly competitive world of molecular biology were being warned away from p53 as a dead-end topic. Even some of the old guard who had been in the field since the discovery of the new gene thought of abandoning this line of research and looking for new challenges.

For nearly a decade p53 research was in the doldrums. But as people began to cotton on to the fact that for several years they had been led up the garden path by mutants, a big question took shape in the collective mind: if p53 was not an oncogene, what else could it be?

CHAPTER SEVEN

A New Angle on Cancer

In which we hear of the revolutionary discovery of tumour suppressors – genes whose job is to protect us from cancer by detecting and eliminating cells with corrupted DNA.

***

Science, my lad, is made up of mistakes, but they are mistakes which it is useful to make, because they lead little by little to the truth.

Jules Verne

In retrospect there were many experiments with p53 that raised questions about its true function. Varda Rotter works in a lab two floors down from Moshe Oren at the Weizmann Institute. Now in her sixties, she's a handsome, stylish, vivacious woman with strong features and short pepper-and-salt hair; she has been involved with p53 since its earliest days, frequently swimming against the tide as her research findings have challenged mainstream thinking. Rotter's passion for her work is obvious: the gene is almost a character in her life. She uses word like 'beautiful' and 'monster' and 'schizophrenic personality' to describe it. And on the walls of her office in Rehovot – among the framed photos of iconic microscope slides and of Varda with colleagues at various workshops – are pictures, drawn by her granddaughter, of p53 depicted as an angel and a devil. She smiles broadly and with obvious affection as she tells an anecdote about how her obsession with her work has been seen by her family. Her daughters were still little when Rotter joined the lab of the Nobel Prize-winning virologist David Baltimore at MIT as a postdoc and the family moved to the US for a while. 'On one occasion they were asked at school what their parents did, and my older daughter answered, "Oh, we have a dad and p53!"

'I do my work with all my emotion; that's just how I am. Even today my head of department said, "Why are you showing me with such enthusiasm such benign data?' I said, "Oh, don't you see, this is a breakthrough." And she said, "*Every day* a breakthrough . . . !"' Rotter laughs as she mimics her colleague's exasperated tone. 'But I think if you do not have this attitude you could not survive,' she adds theatrically.

In 1984 Rotter was making her own clones of p53 and running experiments with a virus that infects blood cells and causes leukaemia. History is ruthless with the runners-up or she too might now be credited with discovering p53, for she had stumbled across the gene herself in her research with a mouse virus and its oncogene, called 'Abelson' (or Abl for short), while working in Baltimore's lab in the late 1970s. Like the others, she had found that her efforts to separate Abl protein from everything else in cancerous cells brought along another protein for the ride with a molecular weight close to 53 kilodaltons. But her findings were overshadowed by Lane, Levine and the rest, who published their discoveries just ahead of hers. In her later work with leukaemia cells, Rotter's findings were very different from most others exploring the function of p53. Far from seeing an over-abundance of the protein, she found none – the gene seemed to have been lost. Researchers in Canada also working with leukaemia viruses found the same thing – most of their blood-cancer cells showed no p53 activity at all.

The two groups published their research at the height of the debate about what p53 does. So how, I wondered, did the rest of community square these findings with the prevailing dogma that over-expression of the protein is the very hallmark of an oncogene? Rationalisation, replied Moshe Oren. 'People knew that leukaemias very often have different biology from solid tumours. So you could say, okay, well, maybe leukaemia's a different story – we don't understand p53 in leukaemia, but we understand that in the typical types of cancer p53 is behaving just as an oncogene should behave.'

Not everyone was ready to be convinced by such an argument, however. In 1982 Lionel Crawford, in whose lab the young David Lane had been doing research when he discovered p53, had published a paper describing his work with breast-cancer patients, where he found that almost one in 10 of the afflicted women had antibodies to p53 in her blood serum. This meant that, in effect, the women's immune systems were preparing to attack one of their bodies' own proteins as if it were a foreign invader. Crawford's observations went largely unnoticed because at the time no one knew what to make of them. But they intrigued Thierry Soussi, still working in Paris in the lab of Pierre and Evelyne May.

Standing back from the jigsaw puzzle to get a better view, Soussi wondered: could the fact that most solid tumours produce an over-abundance of p53 protein, and the fact that in some cases this abundance triggers the immune system to produce antibodies, be related? Might this be an indication that the protein – and the gene that produces it – is mutant and therefore malfunctioning, which is why the immune system sees it as foreign? It was a portentous question because, if this was the case, it would mean going back to the drawing board to discover the true nature of p53.

ALFRED KNUDSON AND THE 'TWO-HIT' HYPOTHESIS

For clues to what this true nature might be we need to rewind the clock to the late 1960s to meet paediatrician and polymath Alfred Knudson, then working at the MD Anderson Cancer Center in Houston, Texas. Here Knudson developed the 'two-hit' hypothesis of tumour formation that was to spark a revolution in cancer research. The two-hit hypothesis is such an important concept in the history of p53, and the information about Knudson the man so sparse, that I decided to seek him out myself to hear his story. I flew into Philadelphia where he lives with his wife, Anna Meadows, a paediatric cancer specialist, on a hot July afternoon in 2012.

Arriving at their penthouse apartment in the city centre, I was ushered into a large, elegant drawing room with thick cream carpets, a feeling of space and light, and views over Philadelphia's narrow tree-lined back streets, with their red-brick terraced houses and tiny gardens. The walls of the Knudsons' apartment are hung with artworks – paintings and collages of such vivid colour and compelling compositions that they immediately draw the eye and invite conversation. But what most caught my attention was a striking, man-sized sculpture in scrap metal of a spindly figure with one huge eye thickly fringed with eyelashes made from nails. The Knudsons had spotted it at one of Philadelphia's outdoor art shows and, so pertinent was it to Alfred's seminal work with retinoblastoma, a childhood tumour of the eye that inspired his 'two-hit hypothesis', that they bought it. 'It was made by the wife of one of Al's scientific colleagues, Zoila Perry,' explained Anna, a petite, pretty woman who exudes energy and a sense of purpose. It appealed to the Knudsons because, viewed from one side, the eye is blue and from the other it's green, seeming to symbolise the 'two hits'. 'Both of us have worked in retinoblastoma – Al in the genetics and me in the clinical aspects including new therapy to save eyes,' commented Anna.

That first evening, the three of us took a bottle of white wine from the fridge and went out into the warm summer evening for a meal at one of the Knudsons' favourite Italian restaurants. Next day, Alfred Knudson and I walked together through the city to the College of Physicians. Here we were shown into the library, a hushed space of dark wood and dusty books behind a heavy door, where we sat down to talk. Knudson is now in his nineties, a spare man of medium height with a mop of white hair, a deep, slow voice and eyes that hold yours unblinkingly as he speaks.

Born in Los Angeles in 1922, he was the first person in his family to go to university, but felt an academic career was his destiny from his earliest days in high school. Then, he mostly imagined himself as a mathematician or physicist, but the current carried him towards medicine instead. He got a place at Caltech (the California Institute of Technology), where he was required to take courses in chemistry and biology as well as physics and maths and says with a deep chuckle, 'I hate to admit how naïve I was – and you'll be amused by this . . . I came to the conclusion that they already knew everything in physics and it didn't seem like a very interesting field to get into!' Biology, on the other hand, excited him with its possibilities and he decided eventually to study genetics, which seemed to combine biology with his love of maths. 'It was obvious there was a lot to be discovered in terms of human genetics, because people were almost totally ignorant about heredity and disease at that time.'

Still at Caltech when America entered World War II in December 1941, Knudson was advised that his best hope of staying in school was to join the forces and apply to study medicine or engineering, which were considered strategic skills and sponsored by the government. 'I didn't think I wanted to do engineering – I had the same objections as with physics, only more so!' He accepted a place at Columbia University Medical School in New York, and left southern California for the first time in his life. He decided to study paediatrics. 'Children have some interesting genetic diseases, and I'd had a course in embryology and thought, oh, here's a *great* field that is way behind the times – there are all kinds of possibilities with this. I thought paediatrics would give me a chance to combine genetics and developmental biology, and it did.'

Knudson's first serious encounter with childhood cancer came during his residency at New York Hospital, when he was required to spend a month at Memorial Sloan Kettering Cancer Center just across the street. 'It had a little unit for children's cancers. There were about 20 patients there. I had seen a child with Wilms' tumour before, and somebody with leukaemia – but to see 20 children with cancer just blew me away . . . I never forgot that.'

Knudson rejoined the army for the Korean War, believing they would draft him anyway, but never saw action: he reckoned they had little use for a paediatrician on the front line. After two years kicking his heels at Fort Raleigh, Kansas, he felt life was passing him by, he told me, and he was anxious to get back into the world of ideas. He returned to Caltech to do a PhD in genetics and biochemistry in 1953, just months after Crick and Watson had cracked the mystery of DNA's structure. The institute, on the crest of the wave of genetic research, was buzzing with intellectual energy. The molecular-biology revolution was gathering momentum.

With his doctorate under his belt, Knudson moved on to the City of Hope Medical Center in Duarte, California, to head a new paediatric unit with a special focus on cancer. But it was at MD Anderson in Houston, which recruited him 10 years later to start a programme in genetics, that he developed his two-hit model of tumour formation. Knudson figured that in order to tease out what was happening at a molecular level in cancer it was best to work with one of the less complicated tumour types. 'To start out studying something like polyposis was kind of hopeless, you know?' (Polyps, he explained, are fleshy outgrowths of normal tissue in the wall of the colon that can eventually turn malignant, and the progression to cancer can take years and follow many different paths.) 'But if a kid can be born with cancer it's about as simple as it can get. That was my thinking.' Retinoblastoma met this criterion; it was the ideal topic for research.

A rare tumour of the retina, or light-detecting cells of the eye, retinoblastoma affects children almost exclusively below the age of five, because it starts in the stem cells of the developing retina that, like the stem cells of all organs of the body, experience an explosion of division and growth during gestation and the early years of life. An early sign of the disease is a milky-white appearance to the pupil of the eye which, left undiagnosed and untreated, as it often is in the developing world, will grow into a grotesque spongy-looking mass of red and white flesh that distorts the child's whole face and will eventually kill.

Knudson did not actually see cases of the disease himself. Indeed, he had given up regular paediatrics and treating patients by this stage to concentrate on genetic research, but he struck lucky. He discovered that two people – a paediatrician in England and an ophthalmologist at MD Anderson – had kept detailed records of retinoblastoma cases they had seen. Poring over this rich repository of data, he observed that the disease ran in families as well as occurring sporadically, and that there were distinct differences between the two sorts of patients. Affected children from families with the disease typically developed cancer at a much younger age than those with no family history. And they tended to have multiple tumours in both eyes, while the sporadic form of the disease typically led to a single tumour in only one eye.

So what did Knudson make of what he was seeing? Something 'outrageously brilliant' in its simplicity and far-reaching implications, said Peter Hall, reviewing the pivotal moments in cancer research nearly four decades later. Here in a nutshell is how Knudson himself explained it to me as we sat together in that quiet library in Philadelphia.

Every gene in the body, except those of the sex cells, ova and sperm, is made up of two copies of itself (or two 'alleles'), one copy inherited from the father and one from the mother. Knudson realised that in those cases of retinoblastoma with a family history of the disease, the gene responsible was inherited – though in the late 1960s no one was yet able to isolate genes or identify them individually. The overwhelming likelihood, he believed, was that only one copy of the gene, one allele – inherited from either mother or father – would be faulty at birth and the disease would occur only if and when the other allele developed a fault. His reasoning was confirmed by the fact that among the familial cases, not all the siblings of an affected child developed tumours, though they would all have inherited the same genes from their parents and been similarly susceptible.

In those cases in his data set with no family history of retinoblastoma, faults had obviously developed in both copies of the gene quite by chance over time – explaining why victims tended to be older than those with familial retinoblastoma and typically to have no more than one tumour.

All this suggested to Knudson that one faulty copy of the gene involved in the development of the retina is not enough on its own to cause tumour growth at that site. The other copy has to develop a fault also. But – and this was the revolutionary suggestion – *until* the normal copy develops a fault, it seems to keep the already faulty copy in check, so that the cell functions perfectly normally. Only when both copies are faulty do the cells start to behave erratically and to develop into a tumour. Put another way, in all the cases Knudson observed, something seemed to be breaking down and the cells involved to be losing their ability to function properly. He proposed that what had happened in these cases was that some kind of brake on proliferation of cells was lost. It was such a simple suggestion, but it swam against the tide of cancer research which, in 1971 when Knudson published his 'two-hit hypothesis', was intensely preoccupied with the driving forces of cancer, the newly discovered oncogenes, which implied a *gain* of function. He was proposing the existence of an equal and opposite force, an 'anti-oncogene' (soon to be renamed a 'tumour suppressor') that allows cancer to develop when it is knocked out – implying a *loss* of function.

Here, the car offers a useful analogy for visualising the forces at work inside cancerous cells, as proposed by Knudson. Think of oncogenes as the accelerator pedals, and tumour suppressors as the brakes: a defective accelerator cable might stick, forcing the car to speed up uncontrollably (a gain of function); brake failure (loss of function) will have a similar effect in that the car won't be able to stop. But the analogy can be taken still further, since in most cars there are two brake systems. If one system fails, you still have the other and generally both sets of brakes have to fail for catastrophe to occur. Likewise, for tumour-suppressor genes to be involved in cancer, it usually requires both copies of the gene, both alleles, to be disabled. This is the central lesson of retinoblastoma, and the essence of Knudson's 'two-hit hypothesis'.

HENRY HARRIS HAS THE SAME IDEA

Knudson had arrived at his theory through mathematical modelling of the data before him. There was, however, more direct experimental evidence to back it up. It came from the laboratory of cancer geneticist Henry Harris, an obstinate (by his own admission) and independent-minded Australian who had been recruited in 1952 by fellow countryman Howard Florey – famous for his collaboration with Alexander Fleming in the discovery of penicillin, for which they won a Nobel Prize – to work with him at Oxford University.

Talking of his move from Sydney to Oxford many years later, Harris told an interviewer, 'I got a telephone call from Hugh Ward, the Professor of Bacteriology [at Sydney], to say that he had Florey in his office and would I like to meet him. I said, "I must be dreaming, you mean *the* Florey?" He said, "Yes, come on over." So I dropped what I was doing and I went over and there was Florey. He looked very much like a moderately successful businessman, but his speech was very laconic, very direct and he said, "Ward tells me that you like doing experiments, Harris, is that right?" I said, "Yes, I quite enjoy myself, break a bit of glassware, make a noise." He said, "Well, how would you like to come to Oxford?" I said, "That's like asking a man in the desert whether he would like a drink."'

This was the era of the virologists whose flood of insights into the deepest workings of the cell so excited the cancer community that their theories of how tumours start – typically through the acquisition of aggressive new powers in the cell – overshadowed all else. 'My reaction to this unanimity of opinion was intransigent disbelief,' wrote Harris in a review for the journal of the Federation of American Societies of Experimental Biology, FASEB. He figured that, with or without the agency of a virus, the rate at which mutations occur naturally in our cells is such that if malign mutations were always dominant – that is, able to override any other operating instructions in the cell – hardly a child would be born without a tumour already forming.

In experiments run by fellow cell biologists Georges Barski and Francine Cornefert that seemed to confirm the virologists' theories of the driving forces in cancer, Harris was struck by something the two scientists had dismissed in the interpretation of their results. Their experiment involved fusing malignant mouse cells with normal mouse cells to see which set of instructions prevailed. When, in due course, a tumour developed, they concluded that the genetic material from the malignant cell had dominated that of the normal cell. The fact that the resulting tumour cells had a depleted number of chromosomes they thought was of no consequence. Harris did not agree. Could it be, he wondered, that in becoming malignant the cells had *lost* genes that might have suppressed cancer, rather than *gained* genes that encouraged malignancy? It was exactly the question posed a few years later in Texas by Alfred Knudson, looking at the evidence from retinoblastoma cases.

Over the next few years, Harris and his colleagues at Oxford – in collaboration with a lab in Stockholm that had the best materials to play with – explored this question by fusing malignant cells with normal cells of various different types. They demonstrated conclusively that for the hybrid cells to produce tumours, something in the DNA had to be lost – something that presumably was suppressing the malignant growth while it was still present. They published their findings in *Nature* in 1969, two years before Knudson's retinoblastoma studies – and well before it was possible to home in on the individual gene or genes that might be responsible.

But Harris and Knudson were up against the limits of technology in proving their theories; they were ahead of their time and their ideas caused barely a ripple in the cancer community.

THE FIRST TUMOUR-SUPPRESSOR GENE IS FOUND

That began to change in the late 1970s when cytologists – scientists who study the structure and function of cells – noticed that in the tumour cells of children with retinoblastoma, chromosome 13 was unusually short: it seemed to be missing a large chunk of DNA. What is more, in those children with a family history of retinoblastoma, all the cells in their bodies had a truncated chromosome 13. It gave researchers a place to look for the offending gene, and suddenly a hotly competitive race was on to find it and clone it. This promised to be the novelty everyone was seeking – something that might explain the many anomalies that were thrown up by their pursuit of oncogenes.

But though the discovery had narrowed the field considerably, finding the retinoblastoma gene remained a Herculean task, for chromosome 13 is a mighty bundle of DNA some 60 million base pairs long. Furthermore, scientists weren't even sure whether they were looking for a single gene or a clutch of genes that normally worked in concert to suppress tumours. They got their answers by an almost impossible stroke of luck. Arriving at Bob Weinberg's lab in the mid-1980s, a young postdoc named Steven Friend announced he wanted to clone the retinoblastoma gene. As Weinberg tells it, he met this request from his new recruit with frank astonishment: '*What?* How on earth are you going to do that? You don't know anything about cloning; nobody knows exactly where it is in chromosome 13.' But Friend was not deterred. 'Don't worry. I'll do it,' he said.

With what Weinberg calls 'irrational enthusiasm – totally irrational and illogical', Friend went ahead. He struck up a collaboration with a doctor working at the nearby Massachusetts Eye and Ear Infirmary, Ted Dryja, whose caseload at the hospital included children with retinoblastoma. Driven by concern for his small patients as well as by intellectual curiosity and the desire to learn something about DNA, Dryja, who had no formal training in molecular biology, had started to do some lab research to try to find out what lay at the root of this dreadful affliction. Focusing on chromosome 13, he had chopped out and cloned small fragments of the DNA. For him, this was just a means of acquiring some basic skills, but these clones created useful probes for investigating the chromosome further. Dryja shared his new tools with Steve Friend and, soon after the young scientist began his search, 'Lo and behold, one of these probes landed right in the middle of the retinoblastoma gene and allowed it to be cloned out,' Weinberg told his audience in the lecture hall at MIT. Spreading his arms wide to emphasise the length of the DNA strand, then stabbing with his finger to indicate the extraordinary landing site of the probe, right on target, Weinberg spoke with a voice slow and deliberate with amazement. 'Now you know how many mega-bases each human chromosome is long; and you know how astronomically unlikely this stroke of luck was – or is. But it happened . . . This is what's termed an "unearned run". It was a terrific finding.'

Steve Friend and Ted Dryja published their story in *Nature* in October 1986 and now the world was listening. A scientist called Webster Cavanee, then a postdoc at the University of Utah, had earlier narrowed the field of search down even further, to a specific region on chromosome 13, and his 1983 paper was the first to confirm that Knudson's two-hit hypothesis was right. On hearing the news that the actual gene had been found and cloned, Cavenee commented, 'I take my hat off to these guys. You can call it luck, but they did the right experiment, an elegant experiment, and it worked. What more do you have to do before they stop calling you lucky and start calling you a good scientist?'

Alfred Knudson, too, was excited. 'I'm delighted this has happened,' he said. 'Before, we could only concoct theories about what the retinoblastoma gene does. Now that we have the gene, we can get to work on the facts.'

This was nothing less than a paradigm shift – a whole new way of looking at tumour formation as a battle of competing forces between oncogenes and tumour suppressors, the accelerators and brakes of our car analogy above. It led also to the recognition, finally, that cancer is altogether an aberration. Oncogenes are not there primarily to drive cancer, and tumour suppressors are not there primarily to suppress cancer; all these genes have regular work to do, including promoting or controlling the growth of cells as part of the endless cycle of building and maintaining our bodies. Only when these vital genes become corrupted and start to malfunction do they acquire the ability to cause cancer.

When, very soon afterwards, p53 was finally revealed as being the same kind of gene as the retinoblastoma gene – a tumour suppressor, and a powerful one at that – it had an electrifying effect on the field. Researchers reacted like a flock of starlings over a winter field, wheeling around to fly in a new direction, and those who had begun to lose faith in p53 and to consider moving on to other things returned to their work with renewed enthusiasm.

CHAPTER EIGHT

p53 Reveals its True Colours

In which we hear of the brilliant work and strokes of luck that showed normal p53 to be a tumour suppressor not an oncogene – its job being to press on the brake rather than the accelerator pedal in cells with damaged DNA.

***

People don't realise that not only can data be wrong in science, it can be misleading. *There isn't such a thing as a hard fact when you're trying to discover something. It's only afterwards that the facts become hard.*

Francis Crick

The evidence that p53 had been miscast as an oncogene and was in fact a tumour *suppressor* had been accumulating in parallel with the work on the retinoblastoma gene. At the forefront of this challenge to received wisdom was Bert Vogelstein, a legendary figure in the p53 story, whose lab has been involved in many of the most important discoveries relating to the gene. On a stunningly hot afternoon in July 2012, I travelled to Baltimore to meet him, climbing the stairs of an elegant office block, all glass and sunlight and potted greenery, overlooking the original old hospital of Johns Hopkins.

Vogelstein's lab is famed almost as much for its fun as for its hard work. For years he headed a rock band, a bunch of musicians from his lab who called themselves Wild Type and who played at scientific conferences and other venues. The band broke up when the drummer's wife died of leukaemia and, with children to care for alone, he had to drop out.

'We started the band mainly to develop *esprit de corps* in the lab. Everybody liked it, scientists liked it – we used to play for scientists. And it was fun!' Vogelstein told me later. 'I think it's important to have outside activities. I certainly encourage everyone who works here to do so. Most people who have looked at creativity have recognised that inspiration often comes in the off moments when you're not focusing on exactly what you're doing.'

Arriving a little early for our meeting, I waited in the lobby and leafed through a photo album I found lying on a low coffee table – pictures of Vogelstein's lab 'family' at conferences and social gatherings, and of Wild Type in their heyday playing gigs. Beside the albums on the coffee table was a copy of *Grant Making for Dummies*. When he emerged from his office, hand outstretched, I was struck by how slight a figure Vogelstein is and by the expression of impish fun on his face as he led the way into his large, cool office and motioned towards a swivel chair. Against the back wall I noticed the keyboard which, he told me, he likes to tinkle on from time to time during the day.

Now in his sixties, Vogelstein has an air of restless energy and as he talked about his life in science his conversation was punctuated by an extraordinary high-pitched laugh that sounded like a cross between mirth and tears. He comes from a long line of rabbis – 13, he believes – but he defied his apparent destiny to study science. He came into cancer research in 1978, at the height of the oncogene craze, setting up his own lab at Johns Hopkins, where he had qualified as a medical doctor and spent a few years on the wards. In his spare time at medical school he had worked in the molecular biology lab of Howard Dintzis and loved it. 'I started doing research with Howard just to learn what research was, and I did it every summer and every chance I could get during the school year – when I had an elective, or at nights and weekends. And then I also learnt how to take care of patients. I found both of them very satisfying, but I found the research more intellectually stimulating. It was a tough choice, because the gratification you get from treating patients is often immediate, whereas the gratification you get from doing research can sometimes take years – and sometimes never comes. But probably the defining moments were when I started taking care of cancer patients.'

Vogelstein's first paediatric case was a little girl named Melissa, just three years old, whose parents brought her to hospital because she was pale and had suddenly become prone to bruising. Vogelstein diagnosed leukaemia. 'It's still scary when I think about it, because my own granddaughter is two and a half.' He broke off for a moment, imagining himself in the shoes of the little girl's parents. 'Out of the blue, bang! One day she's fine and the next day she's got cancer. Her father was a mathematician, a young guy about the same age as I was. He asked me, "Why did this happen to my little girl?" and I had no answer. No one did. Intellectually he was asking, "What's going on? What's the basis for this disease?" And we had no idea, absolutely no idea. I mean there were a hundred different theories, but just no idea . . . Cancer was a total black box! You could throw some things at the child, and some kids even back then were responding – many of them were. But they were poisons, you know? It was a nightmare.'

The experience, indelibly etched in his mind, helped tip the balance in favour of research; doing something useful for humanity is part of his family tradition, and Vogelstein wanted to have a go at solving the mighty puzzle of cancer. His motivation to fight the disease was constantly reinforced by the fact that his first lab at Hopkins was directly above the radiotherapy unit; he and his students had to walk through it to get to work, passing rows of very sick people, many of them in wheelchairs, awaiting treatment. It was impossible not to run up those stairs and start working, he said.

Vogelstein's plan was to join the hunt for genes that might be involved in cancer. 'But I wanted to do it in humans – I think that was part of my medical training. I thought the only way to really understand what was going on in the human disease was to actually study humans.' But his idea was flatly rejected; no one would give him funds for his research, he told me with a peal of his infectious laughter. 'I was told that the only way to get insights into the disease process was to use an experimentally amenable system, which meant mice or worms or fruit flies, because you can manipulate them, or tissue culture or something. That was the paradigm of the time – and to a large extent still is.'

So how did he do his research? 'Well, I don't *know* how I did it, okay? I mean I really didn't have any funding . . . I had to rent my own microscopes and use my personal money, my salary, to do it. For a couple of years we were really broke!'

Funders eventually came on board as Vogelstein's lab began to show interesting results. Their initial effort went into 'fractionating' tumours – that is, separating the cancer cells from the normal cells in the lumps of tissue so that they could distinguish clearly between the DNA of the two. 'Fractionating tumours sounds trivial, but it's anything *but* trivial and it's been a major stumbling block for people,' Vogelstein told me. The common practice was to grind up tumour material to use in experiments, so cross-contamination of cells was always a problem. 'We did it just with a razor blade under a microscope. Each tumour would take us four or five hours. Stan Hamilton was the pathologist who collaborated with us. We would spend hours a day in the pathology lab micro-dissecting these tissues.'

Having developed the tools and perfected the art of isolating and labelling the DNA from human colon cancer – chosen as a research subject because you can almost watch cancer develop from a growing polyp that starts off benign – Vogelstein and his team were ready to look for the genes responsible for the tumours. This is when they learnt the truth about 'wild-type' p53 – by following much the same line of reasoning as Steve Friend and the others pursuing the retinoblastoma gene.

'We had all these tumours from colon-cancer patients, hundreds of them which we micro-dissected and we looked at alterations. We could see losses of whole chromosomes and large parts of chromosomes – but we couldn't find oncogenes that were responsible. So we thought, well, these losses, maybe they represent the losses of tumour-suppressor genes . . . Tumour-suppressor genes had been *hypothesised* to exist, but they'd never been *shown* to exist at the time – this was the mid-80s. They were mythical beasts!'

Suzy Baker, who had joined Vogelstein's lab to study for her PhD, was set the task of seeking out these beasts. It must have looked like a wild-goose chase at the time, but she set to with the enthusiasm of youth and inexperience. As with the retinoblastoma gene (most often represented simply by 'Rb'), Baker knew where to start looking for candidates – on a specific region of chromosome 17 which was missing one copy in at least three-quarters of all colon cancers. Her task was to search through the DNA of the remaining copy for an important gene that was mutant and malfunctioning, thus signifying that both brakes on the mechanism had failed, one through being lost altogether and the other through mutation – the hallmarks of a tumour suppressor.

It seemed coincidental that this stretch of chromosome 17 happened to include among its many genes p53, already labelled in all the literature as an oncogene. The fact that it produced lots of protein in the cells Baker was working with seemed to confirm its designation. But because of the niggling anomalies in recent experiments with the gene, Baker and Vogelstein decided to check it out so that they could eliminate it decisively from their search. Baker carefully selected a cancer cell that had already lost one copy of p53 along with the chunk of chromosome 17; she then isolated the gene from the remaining copy and cloned and sequenced it – still at that time a tedious task that took many months. When she finally got the read-out, Baker was fully expecting the gene to be normal. But to her amazement, she found a mutation.

'I re-checked the sequence at least 10 times before surreptitiously showing the data to Janice Nigro, a fellow graduate student, to be sure that I had not lost my senses in the escalating excitement,' she wrote in a review of the discovery for the journal *Cell Cycle*. 'Knowing that Bert would approach the data with his usual critical rigor and logic ("What's the least interesting explanation for your data?"), I tried to be as calm as possible when I walked into his office and announced, "I found something interesting."'

Talking of the moment some 23 years later, Vogelstein gave a peal of laughter. 'This experiment was not meant to show p53 was there, it was meant to *eliminate* it – so we could look for the *real* tumour suppressor and we wouldn't have to be bothered by this any more! I remember it quite clearly: we were down on Bond Street, which is a supermarket that our lab was in back then because we didn't have any room in the regular hospital. It was actually kinda nice . . . but anyway, it was Friday afternoon about three o'clock and Suzy had sequenced the whole thing, and she came back and said, "Look, there's a change." It was a single change – I think it was a C to T change – and she had sequenced both the normal and the tumour DNA from the same patient to make sure it was somatic.* She was excited, but I was concerned, because most interesting results turn out to be artefacts. And also this particular change is *not* a big change. We expected to see some very pronounced inactivating event. This was the kind of change you could easily think would do *nothing* to the function of the gene, okay?'

Vogelstein and Baker repeated the sequencing of p53 from the same patient's tumour over and over again and confirmed it was a genuine new mutation. Then they did the

* A somatic mutation is a mutation in a mature cell that has occurred spontaneously during the course of life, as opposed to a mutation that is inherited and will be present in all the cells, both normal and cancerous.

same thing with tumours from several other colon-cancer patients, once again sequencing the p53 gene from tumours where one copy had already been lost. And they found the same thing every time: a single small mutation in the gene. 'When we found the first one I was still doubtful,' said Vogelstein. 'When we found two out of two, I was getting pretty excited. And when we'd done more, then I was sure. It's just the statistics . . . The chances of you getting this somatic mutation in this many tumours is minuscule – one in a billion or something.'

For Baker, a young scientist just starting out on her career, this was an extraordinary coup. 'The eureka moment of a discovery is usually drawn out over time as the hypothesis is retested and confirmed, and a scientist slowly becomes more convinced by the accumulating evidence that they have made a meaningful discovery. But as a naïve and enthusiastic graduate student, I truly believed that I had identified the critical tumour-suppressor gene in colorectal cancer at the moment that I found the first mutation. Incredibly, it was correct.'

HOT COMPETITION

Unbeknown to Vogelstein, Arnie Levine and Moshe Oren, working independently in New Jersey and Israel, were hot on the same track. You will remember that after a period of mighty frustration when his lab was scoring blanks with their clone of p53 while others were successfully creating tumours, Levine had realised that his was the only clone of the normal, wild-type p53 and that everyone else had mutants. His relief at explaining the presumed failure of his clone quickly gave way to excited curiosity. The very recent discovery of the retinoblastoma gene (Rb) – the first tumour suppressor – set him wondering if wild-type p53 could be the same thing. He repeated the experiments with his wild-type clone that he had initially read as failures

– mixing it with an even wider selection of known oncogenes, including the two powerful ones found so commonly in tumours, Myc and Ras – and in every case he found that the wild-type p53 stopped transformation. 'It trumped everything,' he told me when I visited him in Princeton, clearly relishing the memory all these years later. 'Every time we got a transformed cell it killed it. That was our first clue that we had a tumour suppressor; we didn't in fact have an oncogene, though the mutant was behaving that way.'

Levine's lab submitted their paper to *Cell* in 1989, around the same time as Vogelstein and Baker submitted theirs to *Science*. But before either set of findings had seen the light of publication, both teams found themselves attending the same conference, at Cold Spring Harbor on the outskirts of New York City, at which both were scheduled to give presentations. 'The speaker before me was Bert Vogelstein. I'd never met him before,' continued Levine. 'And he gets up and he says the following: "We've been sequencing p53 in human tumours and our results with colon cancers suggest it's a tumour-suppressor gene, not an oncogene." I almost fell off my chair, because my talk was coming next and I was going to say the same thing.'

Did you feel cruelly upstaged, I asked Levine? He shook his head, 'No, no, I *loved* it! The reason I loved it was two things: because I knew when I published the fact that it was a tumour-suppressor gene everyone in the field would attack that. Why wouldn't they? There were 10 years' worth of oncogene papers, right? I mean 10 years . . . Everyone is committed to an oncogene. You say it's not an oncogene and you better prove it really isn't! Well suddenly I had Vogelstein on my side. I had a second observation, a confirmation. That was the first thing that made me feel very good about this – because you're always worried that you've made a mistake.

'But secondly – and this is what made me smile and never stop smiling – it was in *humans*. That's what Vogelstein found. We were working on mice all the time. I mean, Moshe had cloned the human gene, but there was *very* little evidence that humans were going to have p53 mutations. I think throughout the 1980s most people thought this was a curiosity: SV40 didn't cause tumours in humans; if this was the way it forms tumours in animals it's a curiosity that is intellectually very satisfying, but its application to humans is probably nil.'

Since the meeting in Levine's office in 1987 at which the light had dawned about the true nature of their clones – that Levine was the only one with a wild-type clone, while all the others were mutants – Oren too had been keen to discover the function of wild-type p53. And he too had decided to repeat the experiments with other known oncogenes, teaming them up with the normal p53 clone and looking more closely at the results, for he knew now that 'nothing' meant 'something' after all. It did not occur to Oren that Levine would have had the same idea and be doing exactly the same experiments in Princeton, so he was unaware at the time of just how hot was the competition.

'We got with wild-type p53 exactly the same kind of results that Bob Weinberg's lab was getting with Rb: you suppress transformation; you inhibit the growth of transformed cells. So against this background it was easy to conclude that normal p53 was behaving like a tumour suppressor. But had it been three years earlier – before the discovery of Rb – I must say frankly that I doubt if we would have interpreted it correctly.'

With the retinoblastoma gene, Rb, they did at least know they were looking for a tumour suppressor, because such an entity had been strongly predicted by the pattern of the disease in children, explained Oren. But with p53 there was no such firm foundation for predicting a new model for cancer, only a bunch of baffling anomalies in the experiments with those early clones. 'It's challenging to be totally innovative conceptually, because you do experiments and so often they lead you to results that are artefacts; they don't lead you anywhere. You really need to have something to grab on to to say, okay, here's something that makes sense because we've already seen something like it before and we know that it was true.'

Excited by his results, Oren, like Vogelstein and Levine, submitted a paper to a journal – *Nature* in his case. His paper was under review when, in a fluky rerun of Levine's experience, Oren attended a small p53 workshop in Gaithersburg, near Washington DC, at which the other two were also scheduled to speak. He had no inkling beforehand of his fellow scientists' discoveries, and was amazed when they all came up with the same story. It was a moment of revelation, he said, because until he heard the evidence from Vogelstein's lab, working with human tumours, Oren had remained sceptical. 'The results were very impressive, very strong, but I wasn't sure whether they were clinically relevant or whether this was just some kind of nice laboratory result without any connection to your cancer,' he commented. 'And so when it all came together it was really explosive. We said, wow, this is the greatest thing that has happened with p53!

'I came home very excited . . . This was still the age of written communication, and when I got home there was a rejection letter from *Nature* waiting for me in the mailbox.' There had been a smile in Oren's voice and his eyes were shining as he recalled the events of the Gaithersburg meeting. Then the rejection. He shook his head and looked down at his desk as though seeing again in his mind's eye the letter from the journal lying in his mailbox, feeling the anticipation of opening it and having the value of his discovery endorsed, and not quite believing what he did see. 'I was so upset I didn't keep the letter, but one of the reviewers, I remember, said something like, "Oren is trying to jump on the bandwagon of tumour suppressors; it's fashionable, but it's wrong. It's very clear that p53 is an oncogene." I don't remember who this reviewer was, and I'm sure they'd not admit to it now.'

The exhilaration Oren had felt at Gaithersburg was on a par with the thrill he had felt at pulling out the first successful clone of p53 five years earlier, and the let-down now was almost unbearable. 'I spoke to Vogelstein on the phone and I remember I cried to him about the injustice.'

Oren's paper was, eventually, published soon after the other two, not in *Nature* but in *PNAS*. Vogelstein's lab followed up with another paper from Baker's fellow graduate student Janice Nigro, to whom the excited Baker had first shown her result. Nigro had looked at a number of other tumour types where one copy of chromosome 17 was also known to be lost and, sure enough, she had found the same thing – mutant p53 on the remaining chromosome. Clearly this was a general phenomenon, not one relevant only to colon cancer.

The observations from Vogelstein's lab 'opened a floodgate', commented David Lane: suddenly researchers everywhere began going through historic samples of tumour tissue stored in hospital pathology labs, some dating back to Victorian times. 'With the tools available one could survey literally thousands of tumours very, very quickly for alterations, and in a pretty explosive period in 1990–91, we and others showed that p53 alterations were the most common genetic change that occurs in human tumours.'

The papers from this period mark one of the most important milestones in the history of p53 and of cancer research in general. Not only was the new paradigm for the growth of tumours – as a malfunction of the accelerator and/or failure of the brakes inside cells – strongly endorsed, but p53 turned out to be extraordinary. While under normal circumstances it acts to protect us from cancer, it is corrupted by mutation in more than half of all human tumours – and a much higher proportion still in some specific types. In lung cancer, for example, p53 is mutated in 70 per cent of cases, while the figure for colon, bladder, ovary, and head and neck cancer is 60 per cent and for non-melanoma skin cancer 80 per cent of cases. What is more, instead of being knocked out altogether by mutation and rendered non-functional, as happens with all other tumour suppressors (around 30 have been discovered and confirmed to date), p53 can sometimes simply change character, taking on new and different roles in the machinery of the cell, as I will recount in a later chapter.

So how exactly does the gene work normally? And what happens to us when it goes wrong? These were the next big questions for the scientists.

CHAPTER NINE

Master Switch

In which we discover that p53 functions by attaching itself to the DNA in a damaged cell and taking control of other genes – switching them on and off as necessary to prevent the cell from multiplying.

***

A true scientist is bored by knowledge; it is the assault on ignorance that motivates him – the mysteries that previous discoveries have revealed.

Matt Ridley

How wild-type p53 works as a tumour suppressor – and what goes wrong to cause cancer – were naturally the next big questions for Vogelstein's lab. But another question was also nagging at him in the weeks and months following the discovery that normal p53 is a tumour suppressor not an oncogene. Exactly *when* in the development of a tumour does the p53 mechanism break down? Vogelstein's team's special interest was colorectal cancer, of which they had abundant material, and they were able to show quite quickly that the mutation of p53 occurs at the transition between benign and malignant. In other words, mutation of the gene allows a relatively harmless growth in the bowel, a polyp, to turn nasty and able to spread; examination of the mass before this point will likely present intact p53. 'That's been shown to be generally true of other systems too,' explained Vogelstein. 'We showed it, for instance, with the brain and the breast . . .'

As for *how* the normal gene works, one of the first vital clues came from outside the p53 field, from the lab of Carol Prives, a biochemist working at Columbia University in New York, who later collaborated with Vogelstein on teasing out the detail, and has since become one of the stars of the p53 community.

Prives was born and raised in Montreal, Canada, the daughter of artists who had grown up poor and were anxious that their children get the opportunities in life they had not had. 'They very badly wanted me to go to college, because they hadn't. But my mother also saw it as a venue to find the proper sort of husband!' Prives, an engaging woman with a mop of dark curly hair, permanently smiling eyes behind wire-rimmed specs and a slight lisp when she speaks, laughed at the memory. 'This is really going to date me, but it was an era when the strongest pressure on a woman was to get married!'

Prives went to McGill University, Montreal, where she did extremely well in psychology. But she soon realised that the observational and equivocal nature of the discipline didn't suit her temperament; she dropped psychology in favour of her second subject, biochemistry, which calls for more deductive reasoning and satisfyingly firm conclusions. For her PhD she went to work with the British-born biochemist Juda Quastel, renowned for his work in fields as diverse as the bacteria of soil and crop yields, mental illness and cancer. Prives was not sure until she visited Quastel at his lab whether to go for cancer or neurobiology, and recounted with typical self-deprecating humour how she reached her decision.

'Juda Quastel was recruited to McGill to head an institute in biological sciences, and what they gave him was an old mansion which was bequeathed by one of the scions of Montreal's department-store families,' she explained when I visited her in New York in the summer of 2012. 'It was actually a gorgeous old house – the dining room was one lab and the living room another, and there were still the mouldings in the ceilings, and the chandeliers. It was a very odd situation. When I met him, his office was on the second floor. There were three floors, and the third floor, honestly, it must have been for the servants originally: there were these tiny little rooms and it was very squirrelly. Prof Quastel said to me, "What would you like to study?" I said, "Well, I'm interested in cancer and I'm interested in the brain." And he said, "Take your pick . . . The brain is on the third floor; cancer's on the second floor." So I said, "I'll do cancer."' Prives began to laugh: 'This is how my wonderful career started off – just because I was too lazy to walk up three flights to those squirrelly little rooms!'

It was many years later that she got involved with p53 and, as with so many others in this field, it was as a consequence of working with the cancer-causing monkey virus SV40. After a spell in New York following her doctoral studies, she moved with her family – husband and twin daughters – to her husband's homeland of Israel, where she spent seven stimulating years in the lab of Ernest Winocour at the Weizmann Institute. It was Prives' first real experience of living abroad, and she loved the true foreignness of the country and the intellectual challenge of her work. 'The Weizmann Institute was a great scientific environment at the time – particularly in this area, because there were some very strong people in SV40. It was one of the top labs. Ernest Winocour, actually, was an esteemed virologist and many people went to learn this virus from him.'

Her immediate predecessor as a postdoc in Winocour's lab was Bob Weinberg, and Prives picked up the reins from him. Weinberg had discovered the messenger RNAs – the recipes for individual proteins – made by SV40. It was Prives' task to translate those recipes into actual proteins in glass dishes in the lab, as a first step to working out how the monkey virus causes cancer in its animal hosts. After seven years at the Weizmann, Prives returned to North America to take up a permanent position at New York's Columbia University, and it was here that she began to turn her attention towards p53.

'Those of us who were involved in SV40 research were all obsessed with the large T antigen,' she explained. 'I still maintain it's one of the most amazing proteins people have ever studied. It's multi-functional beyond belief – really an extraordinary protein. p53 was this protein discovered by several groups that binds to large T antigen and it was a very mysterious entity.' However, this 'piggyback' protein did not seem to excite much interest among Prives' fellow large T antigen fanatics: most of her colleagues were wholly preoccupied at that time – the mid-1980s – with the fantastic insights large T was offering them into how the DNA copying machinery works in animal cells.

Prives realised with some relief that here was an opportunity for her: p53 offered a relatively empty field in which to play. She had spent 1985–6 on sabbatical at one of the three key labs working on DNA replication with large T antigen, and says, 'I left there realising I'd be seriously insane to try to compete with these guys.' By focusing on p53 she would be doing her own thing, and she welcomed the new challenge: at this stage, p53 was still baffling scientists as an oncogene that didn't play by the rules.

INSECT VIRUSES AS FACTORIES FOR FOREIGN PROTEINS

Besides exposing the futility of competing in a field for which she felt ill-equipped, Prives' sabbatical experience had taught her that, in order to understand a protein, you need to figure out how to obtain manageable quantities of the stuff to work with. Here she was fortunate to get to know Lois Miller, a specialist in baculoviruses, which are viruses that exclusively infect arthropods – insects, spiders and crustaceans. Historically, these little scraps of life first appear in ancient Chinese texts describing disease among silkworms, which the virus liquidates into foul-smelling sludge inside their skins. They played a part in the decline of the European silk industry in the late 19th century, and today they are a threat to the farmed shrimp industry. But since the 1980s, baculoviruses have been put to beneficial use in the biological control of insect pests in agriculture. Their potential as environmentally friendly pesticides stimulated intensive study of their molecular biology that revealed another, equally invaluable, property – baculoviruses are able to pump out large quantities of proteins, including proteins encoded by foreign genes which have been artificially stitched into their DNA.

'Somehow as a result of our conversations Lois decided to make baculoviruses expressing both SV40 large T antigen *and* p53 proteins. And she very kindly gave us those viruses,' said Prives. Working with the two proteins together, Prives soon made an important discovery – that p53 could inhibit the monkey-virus protein from triggering the copying machinery of the cells. She published her findings not long after the discovery that p53's normal function is as a tumour suppressor.

'Next thing, I get a call from this person I've never heard of called Bert Vogelstein, who said, "Can you send me some protein, we have some interesting ideas?" So I said, "Sure." I didn't know anything about this guy, but he seemed very friendly.' Later Vogelstein asked Prives if she would make him some baculoviruses engineered to provide protein from various p53 mutants taken from the tumours of cancer patients. He wanted to compare the activity of these proteins with that of the normal p53.

FOOTPRINTS REVEAL p53'S FUNCTION

While Vogelstein was busy in his lab in Baltimore, Prives, 270km (170 miles) to the north in New York, was poised to make a momentous discovery about the activity of p53. It came from the work of a new young postdoc, Jo Bargonetti, who had recently joined her lab. Bargonetti had expressed her interest in figuring out exactly *how* p53 manages to prevent large T antigen from switching on the machinery of DNA replication in cells infected by the monkey virus. 'We sort of told her: don't do that, it's boring. We've already figured that one out; go do something else,' said Prives with a laugh. But Bargonetti was insistent: she wanted to do a 'footprint', she said.

'A footprint is a very elegant kind of experiment that people don't do very often,' explained Prives. It involves mixing the purified protein or proteins under investigation with DNA in a test tube and adding an enzyme to the mixture. The enzyme acts like chemical scissors, chopping the DNA up into fragments. However, if the protein is binding to the DNA, the 'scissors' are unable to cut the ribbon of genetic material at that point. Finally the mixture is run through a gel which organises the DNA fragments into a 'ladder'. If no protein is binding to the DNA, the ladder is intact; if a protein has bound to the DNA, there will be a gap in the ladder. 'Looking at the gel will tell you, if you know how to read the sequence, *exactly* where the protein binds to the DNA,' said Prives.

Here a bit of explanation is needed in order to understand the significance of what Bargonetti was about to uncover. Generally speaking, the job of a protein that binds to DNA – attaches itself to a spot on the double-helix ribbon of genes – is to control the expression of genes in that region; to switch them on and off as appropriate. Proteins that do this are called 'transcription factors' and are, in effect, the conductors of the orchestra of activities within a cell. As the protein that switches on the DNA replication machinery in cells infected with SV40, large T antigen was already known to be a transcription factor. Bargonetti's interest in doing footprint experiments using large T antigen and p53 proteins together in her mixture was, among other things, to show physically how p53 stops large T from attaching to DNA to switch on the machinery. 'We told her, "Okay, do it, if you insist!"

said Prives, who then left the young scientist to her own devices without further thought.

But Bargonetti found more than she had expected. 'She came to me one day and she said, "It's working, but I have this really big problem . . . p53 itself is giving me these patterns on the gel."' Recognising a rare eureka moment in science, Prives responded, 'That's not a problem – you are the luckiest person in the world!' The clear – and totally unexpected – footprint left by p53 in the DNA ladder was the first clue to how the tumour suppressor works, for it too is a transcription factor. And we now know it is an extremely powerful one, sitting at the centre of a network that orchestrates life-and-death signals in every single cell in our bodies.

Bert Vogelstein had also sussed out that p53 was a transcription factor, but by another route. His lab had observed that p53 protein is active in the nucleus – the powerhouse of the cell where the DNA is stored – rather than in the cytoplasm, the body of the cell. Transcription takes place only in the nucleus and many of the proteins active in this site are involved directly or indirectly in controlling the expression of genes. So Vogelstein's team did experiments to see whether p53 bound to DNA – a defining feature of a transcription factor – and they found that it did.

'The other equally important part of the story,' explained Vogelstein, 'is that we weren't simply looking to see what DNA sequences it binds to – and finding that it could not only bind to these sequences but also activate "downstream" genes – we were in all cases comparing the wild type to the mutant forms. And what was satisfying and made us think we were on the right track was that every mutant we looked at was devoid of this binding activity. We had cloned them, right? And this allowed us to test unequivocally whether these mutants were disrupting this function – and every single one, without exception, prevented its binding to DNA. So that convinced us – and I think the rest of the world, too – that it was right.'

Jo Bargonetti had also done footprints using protein from Vogelstein's mutants as a control to her experiments with wild-type p53. 'Jo said, "You know, it's only the wild type that's making these patterns, not these mutants,"' continued Prives, putting her hand to her chest to demonstrate her breathless excitement at the revelation. 'I'm going, *arghh* . . . I'm going to die and go to heaven! I mean, it was just so perfect! . . . Jo's footprints weren't great, they were not very clear cut, but it *was* clear that the wild type had a pattern of recognition of the DNA, and these mutants didn't. That was *absolutely* clear.'

Prives paused to reflect and then said, 'You know, I'd been doing science for many, many years, but I don't think anything I'd done up till then was of anywhere near that significance.'

Prives' lab also tested the footprint findings by putting the proteins through their paces in cell cultures in glass dishes in the lab; they found that wild-type p53 did indeed act as a transcription factor and that mutants did not. Meanwhile, Vogelstein's lab showed that they did the same in living organisms. The two labs, who were by this time talking regularly, published their findings as joint authors in the influential journals *Science* and *Oncogene* in 1991.

But what is the connection between p53's activity as a master switch and its role in protecting us from cancer? How does the tumour suppressor *itself* get switched on? And what does it actually do when this happens? These were the next things to consider.

CHAPTER TEN

'Guardian of the Genome'

In which we discover that p53 protects us from cancer by stopping potentially dangerous cells in their tracks as they attempt to divide, and sending in the repair team to mend the damaged DNA.

***

I worry about p53 a lot. I'm paid to do it, but perhaps we all should, as the correct functioning of this 393-amino-acid nuclear protein is apparently all that lies between us and an early death from cancer.

David Lane

A critical part of the answer to the question of *how* p53 suppresses tumours came from a fellow medic of Vogelstein's at Johns Hopkins Hospital, Michael Kastan, who came to p53 research through his work as a paediatric oncologist. 'People say it must be depressing to treat kids with cancer,' commented Kastan when I spoke to him by phone from North Carolina, where he now lives and works. 'But first of all, kids do much better than adults; we cure 80 per cent of children with cancer, which is amazing. Also, they deal with it much better; and you get to know the families really well, so it was socially a more attractive field for me than other aspects of cancer.'

A single case that seemed to epitomise the combined social appeal and scientific challenge of paediatric oncology for Kastan – and convinced him this was the right field for him – was that of Dora Squires, a little girl with an unusual form of Down's syndrome. So-called 'translocation' Down's meant that, instead of having a complete extra copy of chromosome 21, as is most often the case, Dora had an extra scrap of chromosome 21 that had translocated and attached itself to another chromosome. When Kastan, as a young doctor on the wards, met her she was three years old and already had a long history of cancer.

Dora had been born with leukaemia – a not unusual occurrence in Down's children – which resolved itself without treatment, Kastan told me. She did fine until she was two-and-a-half years old, when she developed a tumour on her face, a rapidly growing sarcoma that responded well and melted away with radiation therapy. But the little girl was soon diagnosed with acute myelogenous leukaemia, the condition for which she was being treated when Kastan appeared on the wards.

'So I come on service and I hear this story, and I say, "Here is a girl with a known chromosomal translocation that in the space of three years has had three different tumours, and no one is saving her blood to try to figure out why." Now this is before the era of knowing about oncogenes, but I said, "We don't know what question to ask now, but if we save her blood samples, some day we'll be able to ask questions about what it is with this translocation that led to this story." Scientifically I found it extraordinarily interesting, and I felt it was a field ripe for discovery.

'But the other piece of the puzzle was that Dora was a typical Down's child: she was very very happy, and she always loved to see us when we came on our rounds. Her father weighed 275kg (600lb) and was too large to sleep on the parent bed in the ward, so he used to sleep on the floor – and she would sleep on his belly. So when we came on our rounds in the morning we'd open the door, she would see us and she would sit up on his belly and put out her arms for hugs from everyone on the team. You can't not melt . . . So that's how I decided to go into paediatric oncology – it was because of Dora Squires.'

Kastan had sandwiched his medical training around a PhD in molecular biology, for which he studied how cells respond when their DNA is damaged. And when, after completing his specialist training in paediatrics at Johns Hopkins, he started his own lab – a modest set-up consisting of himself and an inexperienced young assistant – to do research on the side, this was the topic he was intent on pursuing. His daily experience on the wards had convinced him it was central to cancer biology, he says, and nothing has happened since to change that view.

'We know DNA damage causes cancer, right? We know this from animal models where you can take an animal, treat it with a carcinogen or radiation and cause tumours. We know it from human experience – Hiroshima and Nagasaki, for example, showed us how radiation causes cancer. We know it from exposure to carcinogens in the environment, which is why we have laws about what chemical companies can put in the water. And we know it from familial cancer syndromes, most of which are due to mutations in DNA repair genes.

'So we know DNA damage *causes* cancer. But we also use DNA damage to cure cancer: radiotherapy and most of our chemotherapeutic agents target DNA. And most of the side effects of treating cancer – the hair loss, the bone-marrow suppression, the nausea and vomiting – are because DNA damage is killing normal cells. So from a clinical perspective, or a cancer-biology perspective, DNA damage causes the disease; DNA damage is used to treat the disease; DNA damage is responsible for the side effects of treating the disease . . . It makes it a pretty important phenomenon! As an oncologist, understanding and characterising DNA damage signals is important in every aspect of cancer.'

Kastan's research had always focused on blood cells and what goes wrong with them to cause leukaemia. It was his preference for this cell type over the ones that cause solid tumours, or carcinomas, that enabled him, serendipitously, to discover the key mechanism by which wild-type p53 protects us from cancer – simply because, unlike carcinoma cells, leukaemia cells almost never have mutant p53.

Thus what he observed in his experiments was the activity of normal p53 under a variety of circumstances, whereas he would not have seen anything – because nothing would have happened – if the cells had contained mutant p53. It was serendipity also because p53 was far from his mind, with no place in his research agenda, when he began his experiments to look at DNA damage and repair in cells that are dividing.

The cell cycle, as this dividing process is called, has several phases, and Kastan had been intrigued by a paper he had read from researchers studying yeast – one of nature's simplest, most pared-down organisms, consisting of a single cell – that described how, if its DNA was damaged by radiation, yeast would stop at a 'checkpoint' in its cycle while the DNA was repaired, before carrying on through the cycle. The two researchers had found the gene responsible for this exquisite control of the cell cycle after damage, and this set up a challenge to Kastan: could he identify the genes and proteins that might be doing a similar job in us?

The first thing he did with his leukaemia cells was to bombard them with ionising radiation – which typically causes extreme damage to the DNA by breaking both strands of the double helix – and to take note of the changes in the cell cycle as a consequence. He found that his damaged cells arrested at checkpoints, demonstrating for the first time that what happened to yeast was a general phenomenon, applicable to the human body too. Now he could start asking the questions that really interested him: were there proteins whose level in the cells increased as a result of the damage, indicating that a particular gene or genes had been activated by the event? And if so, which of these genes were responsible for arresting the cells at the checkpoints?

Kastan was pinning his bets on the known oncogenes, and he was surprised to see no changes in these. But he had developed especially sensitive tools for measuring protein levels, and he noticed that the reading for p53 was slightly elevated. This was unusual since, for reasons that will soon become clear, p53 protein is normally present at levels that are barely detectable in cells. Could this slight increase be significant, he wondered?

Indeed it could. As we shall see, Kastan had begun to uncover the mechanism by which p53 suppresses tumours – by halting defective cells in their tracks so that they cannot divide. He was on the brink of a momentous discovery, but it would take time and hard work to tease it out.

FOLLOWING THE CLUES

While Kastan was busy with his initial experiments, a paper came out from Steve Friend – the scientist, you will recall, who had discovered the first-ever tumour suppressor, the retinoblastoma gene, Rb. Friend's paper showed that if you pushed a dividing cell into producing an over-abundance of p53, it came to a temporary halt at a checkpoint named G1. It was a simple observation; Friend did not know whether this ever happened in real life, nor what might trigger an over-abundance of p53, but it made Kastan sit up. Could this be part of the same picture he had observed in his damaged leukaemia cells? Could the damaging event, the ionising radiation, be what activated p53? And could p53 therefore be the protein that was *responsible* for the checkpoint arrest?

'This is when I got Bert Vogelstein involved,' he says. 'You know, I was a clinician who happened to have a very small lab and was doing these cell-cycle studies; Bert had this big machine . . . Since we were at the same institution I knew they were able to sequence p53. So I got on the phone to him one day in between seeing patients, and I said, "Bert, I think we know what p53 is doing. Will you sequence these cell lines for us?"' Kastan described his experiments and his hypothesis to Vogelstein and explained that he wanted to check which of his cells had wild-type p53 and which had mutants, and to compare their activity. 'Bert didn't believe a word of my story!' laughed Kastan. 'But he said sure, he'd have the cells sequenced. And lo and behold, when we tested them, those that had wild-type p53 arrested at the G1 checkpoint after radiation, and those that had mutant p53 didn't . . . I immediately had a sense of how important this might be: all of those experiments told us p53 plays a role in DNA damage responses.'

Kastan published his findings in *Cancer Research* in 1991 hot on the heels of news from Vogelstein's own lab, working in collaboration with Carol Prives, that p53 was a master switch. All of a sudden, scattered pieces of the jigsaw began to fall into place.

THE JIGSAW BEGINS TO TAKE SHAPE

Another of those jigsaw pieces came from Kastan's clinical casebook and involved a rare, inherited, neurodegenerative condition called ataxia telangiectasia, or AT. This devastating disease affects between one in 40,000 and one in 100,000 people worldwide. Children typically start to show signs of AT as toddlers, as it kills brain cells and progressively disrupts their motor co-ordination, affecting everything from walking and balancing to speaking, swallowing and moving the eyes; it generally sees them wheelchair-bound by the time they are in their teens. Treating his young patients with the condition, Kastan knew their risk of cancer was exceptionally high – in fact, 37 to 100 times that of the general population. He knew also that they were especially sensitive to ionising radiation, so medical procedures such as X-rays and CT scans were to be avoided if at all possible. Now, he began to wonder if both phenomena might have something to do with p53. Were the cells of patients with AT able to halt the cycle at the GI checkpoint and activate a DNA repair programme as they should, or was this mechanism defective?

'We had cell lines from patients with AT and it became

clear very early on that p53 did not get induced normally,' Kastan says. 'We had absolutely no idea what the gene was that was missing in these patients. But whatever it was, we realised it was somehow required for the induction of p53 after radiation.'

At the same time as investigating the AT connection, Kastan was collaborating with a scientist at MIT, Tyler Jacks, who had created experimental mice with no p53. Sure enough, thymocytes – important components of the immune system – in Jacks' mice failed to arrest at the G1 checkpoint when bombarded with radiation. Together with Vogelstein, Kastan was also collaborating with a third group, at the National Cancer Institute near Washington DC, who had discovered a collection of genes called GADDs that are directly responsible for arresting growth of cells with damaged DNA (indeed, their name is derived, imaginatively, from Growth Arrest and DNA Damage). The three teams found that GADD 45 was controlled by p53, and was one of the genes switched on by the tumour suppressor to cause arrest at the GI checkpoint that Kastan had first uncovered. Very soon, Vogelstein found another gene, p21, involved in the same event and also controlled directly by p53.

The picture that emerged of p53 from these disparate bits of research was of a master switch at the hub of a communication network within cells. Its job is to respond to incoming signals indicating DNA damage by recruiting the relevant genes 'downstream' to halt growth of the cell pending future decisions about its fate. In this way cells with scrambled DNA that might threaten the organism are disabled.

It was this picture that the researchers described in a paper they published together in *Cell* in 1992 and that Kastan says was 'the most fun thing I ever did in my scientific career'. Just before the paper came out, he attended his first big p53 meeting, hosted that year by Moshe Oren and Varda Rotter in Israel. Oren had seen Kastan's original paper on checkpoint arrest following radiation and been sufficiently excited to invite the American to speak at the plenary session – to the full, august gathering of the p53 community. 'What was so much fun was that I was a *total* unknown in the p53 field,' said Kastan. 'I go to this meeting; I get up to the podium and give this talk about this whole signal transduction pathway the day the paper was published in *Cell*.' No one had seen the data before, and it had a powerful effect on the audience.

'I was a nobody with a no-technology lab,' he continued, 'but I just happened to ask an important question because I read the literature carefully. And I asked it at the right time, with the right techniques and in the right cell type.'

RARE DEGENERATIVE DISEASE HOLDS THE KEY

Not everyone was ready to accept Kastan's model entirely. The fuzziest part of the picture at that stage, in 1992, was the connection with ataxia telangiectasia. No one knew what the missing element was in these patients that made them so sensitive to radiation; they knew only that, in normal circumstances, it was essential for signalling to p53 that the DNA was dangerously damaged and for turning the whole damage-response system on. Things became clearer when, after a Herculean effort by 30 international scientists and hot competition between the labs to find the gene or genes responsible, a team led by Yossi Shiloh at Tel Aviv University announced success in 1995.

The single-minded search for the AT gene took more than 15 years of his life, Shiloh told me when I spoke to him over the phone from New York, where he was on sabbatical in 2012. It began when his mentor at university, Professor Maimon Cohen, suggested that the young scientist join him on a field trip to a small village in southern Israel; there they would meet a family of Moroccan Jewish origin afflicted with ataxia telangiectasia. Shiloh had recently completed his Masters degree and was casting around for a topic for his PhD thesis. 'Professor Cohen had a hidden agenda – to interest me in AT,' he said. 'It worked very well because when I saw those patients I decided almost on the spot that this was an important problem to work on. First, because it's an extreme human tragedy and at that time it was an "orphan disease" – no one cared much about these rare diseases with long names. And second, it was clear that understanding AT would have broad ramifications in many areas of medicine – neurology, immunology, genetic predisposition to cancer and whatnot – because AT is like a microcosm of medicine, it involves so many systems in the human body.'

Shiloh had no illusions about how difficult it would be to find a common cause for such diverse symptoms – and for many years the consensus among AT researchers was that there were four distinct types of the disease and probably at least four different genes responsible. The first breakthrough – what Shiloh identifies as the starting gun for the race to find the genes – came from Richard Gatti at the University of California in Los Angeles, whose study population was the Amish people of Ohio. In 1988, Gatti had managed to localise the gene responsible for AT to a region on chromosome 11, homing in on this stretch of DNA through a technique called linkage analysis, which looks for genetic markers – small strips of DNA with unusual 'spelling' dotted along the genome that are consistently present in people with a particular genetic disease, and never found in healthy individuals. The researchers then use statistics to suggest which marker or markers is closest to the target gene. This narrows the search area, but finding the actual gene is still akin to looking for a person's house when you have only the name of the city in which they live to go on, and it was another eight years before Shiloh and his team managed to achieve their aim.

'When I look back I'm surprised yet again that for eight years the *entire* lab was working on that one project . . .' he said. 'You know, scientists are very individual . . . Today we still work on AT, but every student in the lab has his or her own project. At that time the entire lab, several generations of students and postdocs, was focused on just fishing out genes from that region of chromosome 11, analysing them, cloning them.'

Today, thanks to the Human Genome Project and the wealth of data about genes and sequences available at the click of a computer mouse, such an exercise is relatively straightforward. But in the mid-1990s it was slow and labour-intensive, and relied on close co-operation with the AT-affected families whose personal DNA was the lifeblood of the research. Among the hundreds of genes Shiloh's team cloned was one that specially caught their attention because it was unusually long – so long in fact that they had to repeat the cloning exercise a number of times to convince themselves it was real. Clearly this was the recipe for a huge protein – and one, they soon discovered, that had the hallmarks of a 'signalling' protein responsible for sending messages within the cell.

Shiloh remembers the day they realised this was what everyone had been looking for. 'I had been teaching and when I came back to the lab from my class my student was holding a Southern blot* in her hand. She said to me, and I remember her words clearly, "There is something odd about this gene in this family." This was one of our Palestinian Arab families. I looked at the blot and it was clear that a big portion of that specific gene was deleted in that family. It was a very dramatic result. Of course my heart skipped several beats, but I said to her as calmly and quietly as possible, "This indeed looks interesting, there might be something here. Why don't you repeat the experiment with DNA samples from the entire family and additional controls?"'

She did so and the conclusion was inescapable: here was the gene whose corruption was the cause of the disease Shiloh's team were seeing in all their AT patients. It was a

* The read-out of a procedure that looks at isolated bits of DNA.

time of high tension, recalls Shiloh. The race to find the gene was at its peak, with frequent rumours in the air that someone or other had succeeded, and the temptation to publish his lab's results immediately was heavy. But he had a hunch that there might in the end be just one gene – not the four that everyone supposed – responsible for the different manifestations of ataxia telangiectasia, and it would take time to prove it. Someone else might get there first, but after intense discussion among themselves everyone in his lab agreed to hold off announcing their results until they had tested their hypothesis. It was a nail-biting time, but the gamble paid off: AT is indeed caused by defects in a single gene, which the international consortium named ATM, short for ataxia telangiectasia mutated.

This was the missing detail in Kastan's picture of the DNA damage response: in time he and others were able to show how the signals are passed down the line from ATM, which first senses the broken strands of DNA, to p53, which then throws the relevant genetic switches to halt the division of the cell. This was biochemical proof of the mechanism, and it finally convinced the doubters that p53's response to DNA damage is at the heart of its action as a tumour suppressor.

'You know, you can't overstate the importance of what Yossi did in cloning the AT gene,' commented Kastan. 'He will be somewhat humble in telling it, but people were searching for that gene for 20 years – including him – and it made such an impact . . . It really opened up the whole DNA damage-signalling field. Yossi is a fastidious scientist and it's because of that fastidious approach that they got to that point.

'He flew to Baltimore to tell me he had the gene clone, and I remember very distinctly, he was sitting in my living room, saying, "Okay, we got the gene, and we're calling it ATM for 'ataxia telangiectasia mutated'." I looked at him and I said, "Well, that sounds great, but you know ATM has another meaning in the US?" And I explained to him about these new automated teller machines. His face dropped, and I said, "Don't worry, Yossi, people will know that's where the money's at in the signalling pathway!" And it's been true. The field just exploded at that point.'

MULTIPLE STRESSES, MULTIPLE RESPONSES

In labs everywhere, researchers began testing the model, and evidence soon mounted that many more insults to the DNA – as well as more subtle stresses on the cellular machinery – can trigger the p53 response to halt the cell cycle. The increasingly long list includes UV radiation from sunlight, chemicals in the environment, and activated oncogenes, as well as natural ageing and dangerously low levels of oxygen and essential nutrients like glucose in the cell. Importantly, each stressor has its own characteristic pathway – from the protein that sends out the first alarm signal that all is not ideal for the division of the cell, thus triggering the response, to the range of genes that p53 switches on. But they all have the same effect of preventing potentially harmful mutations from being passed on from one generation of cells to the next.

The frenzy of activity among researchers was fuelled also by revelations that, just as there are many different stressors that can trigger p53, there is also a variety of outcomes to the response. Besides inducing a temporary halt in a dividing cell while DNA damage is repaired, p53 can induce a state of permanent arrest, called senescence. And under certain circumstances, it will instruct a seriously damaged cell to commit suicide – a process that many people feel is the most important weapon in its armoury.

In July 1992, David Lane, p53's co-discoverer, pulled all the information together from widely scattered publications in a review for *Nature* in which he dubbed p53 'the guardian of the genome' – essentially, the policeman in our cells taking action to clear dangerous individuals from the scene. As a reflection of what many people were thinking, it was neat; but as a statement from a scientist it was unusually bold. 'In a sense it was sticking my neck on the block,' said Lane with a mischievous chuckle. 'You write a scientific paper and you say: it's not unreasonable to speculate . . . But in this I said: this is how it works! Then everyone thinks, well there's a challenge! But is it true? Is it not true? Not everyone believes it even now, but it provoked debate, which is what it was intended to do. That's very important to the progression of science.'

AHEAD OF HIS TIME

There is a poignant footnote to this story. It involves a young scientist called Warren Maltzman, who worked briefly as a postdoc in Arnie Levine's lab in the early 1980s, before moving on to Rutgers, the State University of New Jersey. Maltzman's doctoral research at Stanford had focused on how cells repair damaged DNA, and when he joined Levine's team he became involved, naturally, in p53. At Rutgers the two fields came together when Maltzman observed that in normal, non-cancerous cells subjected to UV radiation (as in sunlight), the levels of p53 shot up. He published his findings in the journal *Molecular and Cell Biology* in 1984. 'At that stage,' says Levine, 'we didn't know p53 was a tumour suppressor; we didn't know what it meant that the level went up, and so his paper was *roundly* ignored. Had everybody picked it up, we'd have known p53 was involved in DNA damage and repair responses right away; we might have found that it transcribes genes . . . But . . . the time was not ready for anybody to make sense of it.' Despite a good reference from Levine when he applied subsequently for a research post, Maltzman's academic career faltered and he went into industry. 'I feel badly about that because this man made a contribution whose time had not come . . . In many ways it's the human story of science,' mused Levine.

CHAPTER ELEVEN

Of Autumn Leaves and Cell Death

In which we discover that another, even more powerful strategy p53 uses to suppress tumours is to drive damaged cells to commit suicide.

***

In science it happens every few years that something till then held to be in error suddenly revolutionises the field, or that some dim and disdained idea becomes the ruler of a new realm of thought.

Robert Musil

Paradoxically, one of the most important and dynamic topics today in biology – the science of life and living organisms – is death. Programmed cell death, or apoptosis, to be more precise, which rivals p53 for the number of scientific papers it has generated. But this is not death as most of us know it – a process of decay and putrefaction as the cells in our tissues rupture and spill their contents, to be colonised by bacteria that release bad smells. It is not the death that produces pus in wounds. Necrosis is typically caused by random, traumatic injury and the spilled contents of ruptured cells can cause damage to surrounding tissues, seen as inflammation. Apoptosis, on the other hand, is an integral part of the programme of life – a recycling process in which the cell membrane remains intact while the contents are systematically chopped up and repackaged before being engulfed by phagocytes, the scavenger cells of the immune system, or swallowed by neighbouring cells.

Apoptosis is a process of shrinkage and quiet dispatch. It is unmessy and unseen. For decades it was the territory of embryologists and entomologists, for this is what sculpts our bodies in the womb, removing the web of skin between fingers and toes, hollowing out tubes, shaping organs and building our brains. It is what makes the tadpole's tail shrink as it grows into a frog. And it is part of the process of metamorphosis, whereby a caterpillar turns into a butterfly or moth in the chrysalis, or a nymph into a dragonfly. Indeed it was an entomologist, Richard Lockshin, who coined the term 'programmed cell death' – a decade before it was given the alternative name of apoptosis – to underline the fact that here was a process controlled by the genes, with a beneficial role in biology, not the result of accidental or destructive forces.

Rick Lockshin was one of the earliest biologists to study the phenomenon and became a founder member of the International Cell Death Society and a leading light in the community. He came to the topic as a result of an interest in metamorphosis that developed during his undergraduate years as a biology student at Harvard, when he was given the opportunity to 'hang out' in the lab of entomologist Carroll Williams, and work as a dishwasher and lab technician. 'Williams was one of the world's experts on insect hormones, and he had brought the custom of afternoon tea back from a sabbatical in England. I therefore spent many afternoons listening in fascination to discussions about the mechanisms of insect metamorphosis,' he told an interviewer for the journal *Cell Death and Differentiation* on the occasion of his 70th birthday in 2008.

Williams subsequently became Lockshin's PhD supervisor and programmed cell death the topic of Lockshin's thesis. His research was given a huge boost when Williams, on a trip to Japan, found that moth pupae were selling for a very good price and ordered 20,000 to be shipped to his lab at Harvard. 'When they arrived, he was horrified to realise that they had all initiated metamorphosis during the voyage,' said Lockshin. 'They were going to be nearly useless to almost everybody but me, as long as I was willing to work non-stop, and I was. For a brief time I had more material than I could have ever dreamed of having.'

Programmed cell death began to emerge into the mainstream of biology with the work of three pathologists, John Kerr, Andrew Wyllie and Alastair Currie, who came together in 1971 at Aberdeen University in Scotland, where Kerr was spending a sabbatical year away from his home town of Brisbane, Australia. Kerr had long been intrigued by cell death, having first noticed the phenomenon in London in 1962 while doing research for his PhD, which involved examining the effect on rat livers of cutting off the main blood supply. He could see clear evidence of necrosis in large patches of the livers, which showed all the characteristics of degeneration under the microscope. But gazing down the eyepiece at the wafer-thin slivers of liver, he saw something else, too – single cells scattered sparsely through the living tissue; small round blobs of cytoplasm speckled with fragments of DNA. This was death, too, but without the degeneration, or the inflammation of the surrounding tissue. Unaware of the literature in the insect and developmental-biology fields, he called what he saw 'shrinkage necrosis' because of its apparent role in atrophy of the damaged livers.

Back in Brisbane, in the Pathology Department of the University of Queensland, Kerr began studying the process more closely under an electron microscope. Soon he was examining tissues other than liver and finding the same thing – notably in sections of skin cancer and other tumour types. He and his colleagues concluded that programmed cell death must be responsible for the shrinkage of tumours after treatment, and sometimes spontaneous shrinkage too.

Kerr's series of time-staggered electron micrographs published in a journal caught the eye of Alastair Currie, Professor of Pathology at Aberdeen. Currie was seeing something similar in work with a young PhD student, Andrew

Wyllie, when they treated the adrenal glands of their lab rats with steroids, causing atrophy. Wyllie, who went on to make important contributions to p53 research, had come under the influence of the energetic and generous-spirited Currie as a medical student and been taken under his wing to study for a PhD. Currie died in 1994, but when I met up with Wyllie on the fringe of a pathology conference in Sheffield, he told me, 'Alastair was interested in everybody and knew everyone's names. He took an interest in individuals – and he took an interest in me.' As we sat drinking coffee in a side room, he recalled with affection the 'great gladiatorial discussions' between Currie and his students which the professor, a man with a sharp mind and mischievous sense of humour, clearly loved.

Before accepting the post in Aberdeen – which brought him and his large family back to their beloved Scotland – Currie had spent three years as Head of Pathology at the Imperial Cancer Research Fund in London, and one of the questions that had intrigued him from the start was how tumours shrink. With no obvious signs of death, it must be some kind of cell 'drop-out' process, he concluded, and this is what he set his young PhD student to investigate.

'I began to wonder in a very immature way if the regression was part of a process which had wider significance,' Wyllie told me. 'A lot of the things that tumours do are kind of caricatures of things that normal cells do. And if normal cells go through cycles of death and birth, then maybe the regression of tumours has something to do with that. These were ideas that were floating around in the ether, but it was difficult to design experiments to take them much further.'

The experiments with the adrenal gland were set up to test normal physiological processes and it was here that Wyllie and Currie began to see the single scattered cells that had intrigued Kerr. But the two Aberdeen pathologists were working with ordinary light microscopes which were incapable of showing the little round blobs in any detail, and

Currie was excited by what he saw in Kerr's high-resolution images. He managed to combine a visit to his daughter working in Australia with a spell as a visiting professor in Brisbane, where he made a point of meeting Kerr, and he suggested that the Australian spend his upcoming sabbatical in Aberdeen.

'Before he went, Alastair drew my attention to some beautiful papers of John Kerr's in the *Journal of Pathology*, and I have to say that I didn't catch on initially.' Wyllie, a slight, bespectacled Scotsman now in his late sixties, who speaks softly, precisely and with his whole body, gave a gleeful laugh as he recalled his first reaction to what turned out to be the start of something truly momentous: collaborative work between the three men that would uncover one of the most fundamental processes in biology and change forever the way cancer was perceived.

On arrival in Aberdeen to start his sabbatical, Kerr looked at the adrenal-gland tissues under his electron microscope and confirmed the presence of 'shrinkage necrosis' with identical characteristics to what he had seen in his experiments back home. Wyllie had found the same phenomenon in breast tumours in rats that shrank when the rats' ovaries were removed, depriving the tumours of the hormones on which they were dependent. Hearing about their work, Allison Crawford, a developmental biologist also doing a PhD in the Pathology Department at Aberdeen, drew their attention to the extensive literature on programmed cell death in the developing embryo. It is a sign of how single-minded and narrowly focused scientific research can be that none of them had been aware of this rich body of knowledge before. But now they knew that what they themselves had seen in a variety of tissues and under a variety of cellular conditions, both normal and pathological, was a natural process with a role to play in many aspects of life – a process essential and complementary to mitosis, or cell division, in regulating the population of cells in an organism by clearing out old, damaged or excess cells as new ones are made.

According to Wyllie, Alastair Currie was troubled by the name 'shrinkage necrosis', which didn't distinguish it sufficiently from the processes of putrefaction. And 'programmed cell death' seemed to suggest it was a developmental programme and nothing more. 'Without question, the first description of the phenomenon, the first proper analysis of it, was John's. But I think it was Alastair's vision to emphasise the stereotypical quality of a process of cell death which was different from necrosis,' he told me. 'And then the funny bit of the story . . . This could only happen somewhere like Aberdeen, a small university, an outgoing individual meeting other professors at lunch . . . Alastair met the Professor of Greek and Latin, James Cormack, and he asked him to suggest a term. If the term rhymed with "mitosis" it would be kind of handy as well.' Cormack proposed 'apoptosis'. It was a word from ancient Greek poetry describing the dropping of leaves in autumn, and had been used in a medical context already – by the Greek physician Galen almost 2,000 years previously, to describe the sloughing of scabs from wounds.

Kerr, Wyllie and Currie introduced the term to the wider world in August 1972 in a paper in the *British Journal of Cancer.* 'Apoptosis: a basic biological phenomenon with wide-ranging implications in tissue kinetics' has clocked up a huge number of citations in the scientific literature over the years, but at the time it was received with stunning indifference. 'The stone just dropped into the well,' recalled Wyllie, beginning to laugh. 'In fact, *The Lancet*, in their Christmas edition, had a crossword quiz and one of the questions was: what is apoptosis? We were the joke of biology! "These guys are studying *death*, ho, ho!"'

The trio was not deterred, however, and, together with a handful of others ready to brave being called nuts, they worked away beyond the limelight to uncover the mechanics of apoptosis and to chip away gradually at the single-minded obsession with cell growth that gripped the mainstream cancer community at that time. But it was the discovery of a link between the p53 DNA damage response and apoptosis – or 'cell suicide', as some like to call it – many years later that finally convinced the sceptics of the wide significance of this phenomenon and revolutionised the thinking about cancer.

Thus far, the prevailing view had been that cancer was a disease of anarchic growth. Apoptosis suggested a complementary model of tumour formation: one in which cells that grow at a normal rate fail to die at the appropriate time. Filling a bath with water is a good analogy here. You can fill it up by turning on the tap faster than it can run down the plughole (proliferation) or you can leave the tap at the normal speed, but put the plug in to stop the water draining away (blocking cell death). You end up with the same effect by two different routes.

Serendipity played a role in this radical change of perspective about cancer, as it has in so many important scientific discoveries (think penicillin and the little dish of cultured bacteria Alexander Fleming accidentally left uncovered on his lab bench, allowing it to become contaminated with an antibiotic mould). In the case of p53 the serendipitous events occurred in the lab of Moshe Oren in Israel. The year was 1990. Oren and his team were investigating the activity of p53 mutants known to be oncogenic, trying to find out how they contribute to malignancy, when they were asked to move their lab up one floor at the Weizmann Institute. They were a small operation at the time and they soon had their equipment, including their two incubators, up and running in the new setting and were able to carry on with their experiments. But they began to notice that cells in one of the incubators were no longer flourishing as they had downstairs, and as they continued to do in the other incubator, even though both offered an identical environment and were set at 37°C (99°F). What could be wrong?

Cells not growing are bad news in a lab because there is 'nothing' to study, and Oren's group tried all the usual tricks, like changing the culture medium in the dishes, to coax the sluggish cells back to life. 'But after two months of doing the obvious things and still having problems, we sat together and said, "Okay, what's going on here?"' explained Oren, when I visited him at the Weizmann. 'We realised that only some cells – and only in that incubator – were not growing well. And when we looked more closely, we realised that it was only the cells that had a particular mutant that were affected.' Clearly something about the incubator was making just the cells with that p53 mutant unhappy, and the team decided eventually to check the temperature. 'I was kind of ashamed that we didn't do this earlier. When I was a student I was instructed that you always keep a flask of water in your incubator with a thermometer in, and don't just trust the digital display. And once we did that we realised that this incubator was about 33.5°C (92°F) instead of 37°C (99°F).'

Someone had obviously bumped the thermostat during the removal, and in so doing had revealed an invaluable property of the mutant: it was temperature sensitive. Though the full implications of this took some time to sink in and to test, what Oren eventually discovered was that the mutant behaved like a regular oncogene at 37°C (99°F), the temperature typically set for lab experiments, and like a tumour suppressor – that is, like wild-type, non-mutant p53 – below 34°C (93°F).

Oren was familiar with temperature-sensitive mutants in virology and knew that what he had stumbled across here – in mammalian genes, not viruses, this time – was a very valuable tool; a potential gold mine for p53 research. It meant that scientists could put the mutant into a variety of cell types and watch its activity, first as a regular oncogene helping to drive malignancy; then they could reduce the temperature in the incubator to see how the cancerous cells reacted to the presence of wild-type p53 as the leopard changed its spots. What is more, they could track the reaction over time from Minute Zero.

Oren and his team tested their switchable mutant in a wide variety of cells, and found that at low temperatures, when it was behaving as wild-type p53, it inhibited cell division in those that were damaged – not a new insight by that time, but a confirming one. But they were especially interested to know what would happen in mouse leukaemia cells, which typically have no active p53 at all, if they switched on the wild-type behaviour. So they introduced their temperature-sensitive mutant to a dish of cells given them by Leo Sachs, a leukaemia expert in another lab at the Weizmann, and dropped the temperature in the incubator to 32°C (90°F).

The postdoc given the experiment to do was hoping to see something interesting, but when she returned to check her cells she found to her dismay that they were all dead. 'Usually when you see cells that are all dead you don't think about *interesting* possibilities,' commented Oren. 'You think you've done something wrong and it just means you have to do the experiment again – which she did, and again it repeated. After about two weeks of repeating the experiment, it was clear there was nothing wrong with the way the experiment was done, the cells were just *dying*.'

Oren was quick to realise this was interesting and important, though he didn't know how to interpret it. So he took his data along the corridor to Leo Sachs for comment. Sachs suggested he consider apoptosis – a process Oren had never heard of – and directed him to the still sparse literature on the subject, including the papers by Kerr, Wyllie and Currie. Intrigued, Oren stained the dead cells in his dishes to make them stand out and put them under the microscope. They looked exactly like the textbook images of apoptosis, and further experiments to confirm their findings all pointed to the same thing: 'p53 was killing cancer cells by apoptosis, and we were very excited about that.'

Oren was particularly keen to share his discovery with his old friend and mentor, Arnie Levine, who chuckled as he told me the story of their discussion when I visited him at Princeton. It was on the fringe of a meeting in Vienna in April 1991, where Oren first presented his findings in public and, full of anticipation, approached his former teacher in one of the coffee breaks. 'I have to admit to making an error of judgement!' said Levine. 'Moshe and I are very close because he was my postdoc. Moshe shows me the data about apoptosis, and he says, "Well what do you think? Nice story!" I said, "I don't know if this is going to go anywhere." Moshe looked at me like, oh, that's not good! But that's okay, he's a brilliant guy and he goes on and shows that it is important, and that it's central and so forth. But I'm always amused by the fact that my first response was, "I can't figure out why this would be important to anybody!"

'It goes to show, you know, that you get a mindset about something. You hope that as a scientist you have a completely open mind about things, but of course you get committed to an idea, and you're willing to run with that idea, and that's what makes you work hard on it. But it starts to exclude other ideas, right? And that's just life! That's the way science works.' (It also goes to show just how strong was the prejudice against death as a relevant topic for biologists that, 20 years after Kerr, Wyllie and Currie's paper about apoptosis, some of the most eminent scientists were still so ready to dismiss it.)

Oren's team was the first to demonstrate that p53 can promote apoptosis, but the setting of their experiments and the way they activated p53 in their cultured cells were artificial. The big question was where and when does this happen in real life? It was a question already being explored by scientists working with transgenic mice on both sides of the Atlantic.

CHAPTER TWELVE

Of Mice and Men

In which we hear about experiments with genetically engineered mice to test the activity of p53 in real life against what researchers see in their Petri dishes in the lab. And we learn, too, that the dreadful side effects of conventional chemo- and radiotherapy may be avoidable.

***

Science is helplessly opportunistic; it can pursue only the paths opened by technique.

Horace Freeland Judson

In the long history of p53, huge amounts of data have been generated by scientists poring over little scraps of tissue and clusters of cells in test tubes and Petri dishes – specimens that have been coaxed and manipulated in super-controlled environments. 'These systems are easy and convenient, but they're not the real world,' says David Lane, sounding a note of caution. 'The more I look at p53, the more I realise that in the real world it's operating at a very different level and in a different sort of way.' Tissue culture itself puts cells under stress and p53 into a state of alert, he says, and, rather than studying the difference between active and inactive protein, what most researchers are in fact studying is the difference between very active and moderately active protein. Experiments using animal models tell a story that's different and a lot more subtle.

Recognition of this fact lies behind one of the legendary stories of p53 research, and it involves David Lane and his friend and colleague Peter Hall, both working at Dundee University at the time. The year was 1992. The story goes that the two scientists had been sharing a pint in a local pub at the end of a busy day and mulling over the crucial question of whether or not p53 responds to cellular stress in real life, as it does in tissue culture in the lab. They knew others were asking the same question and that competition to find answers was hot. They knew, too, that they faced a forest of paperwork to obtain Home Office permission for animal experiments, and their frustration at the prospect of the inevitable delay was intense. Then Hall had an idea: why not conduct the experiment on themselves? Without hesitation, he volunteered to be the guinea pig, and the two began to make plans. Telling me the story some years later, Hall said with his characteristic note of defiance that he and Lane knew they risked incurring the wrath of the authorities for not following standard procedure, but they were too fired up at that point to care.

The experiment involved subjecting Hall's arm to radiation from a sun lamp – 'equivalent to 20 minutes on a Greek beach' – and taking a series of time-staggered skin biopsies to watch the activity of p53. 'We reckoned that if this gene does respond to stress in living organisms, we should see the accumulation of p53 protein in the cells in my radiated skin. And that's exactly what we did see,' said Hall, rolling up his sleeve to reveal nine neat scars. 'We did the experiment on me because we wanted quick results . . . The scars all got infected,' he laughed, 'but the experiment worked brilliantly, and it moved the field on considerably.'

Such maverick experiments notwithstanding, yeast, worms and fruit flies have taught us a great deal about how cells work. But for insights into the workings of more complex organisms like ourselves – with organs and skeletons, circulating blood and immune systems – the animal model of choice is the mouse. Similar to us, mice have around 23,000 genes, almost all of which have counterparts in our own DNA. Furthermore, mice are cheap to maintain; they breed fast, producing a new litter roughly every nine weeks; and their genomes are relatively easy to manipulate.

For decades, scientists used selective breeding techniques to produce mice with desired genetic traits. Or they blasted their DNA with chemicals known to produce specific mutations: a process known as 'chemical mutagenesis'. Then in 1989 came the birth of the first transgenic mouse, created using a sophisticated technology called 'homologous recombination'. Such mice provided a new 'precision tool' that changed everything, and homologous recombination won its developers, Mario Capecchi and Oliver Smithies, both working in the US, the 2007 Nobel Prize for Medicine. They shared the prize with a Briton, Martin Evans, who was the first person to isolate the embryonic stem cells from which transgenic mice are created.

The story goes that Evans was on a month's visit to the US, where he had gone to learn some new technological tricks at the Whitehead Institute in Cambridge, Massachusetts. With so little time for his mission, he was determined not to be sidetracked into giving lectures or meeting new people. He didn't even want to speak to anyone outside the lab. Then he got a phone call from Smithies, a fellow Brit who had left for the US many years earlier. Smithies was eager to learn more about Evans's embryonic stem cells, which were so vital to his own research goals. 'I remember to this day, I said to him, "Oliver, you are the only person who I will come and visit . . ."' Evans told an interviewer for the Nobel Committee. And he turned up the following weekend at Smithies' place with a flask of the cells in his pocket.

Homologous recombination – more descriptively known as 'gene targeting' – exploits the cell's natural propensity for repairing breaks in its DNA by stitching in little pieces of matching DNA taken from another chromosome. In gene targeting, scientists insert into the cell a foreign piece of DNA carrying the desired genes, and they rely on it to find the appropriate place (where it recognises a matching sequence of genes) to insert itself into the host DNA, in this case kicking out the original sequence.

Over the decades, this method has been used to create many thousands of mice precisely engineered to model human conditions and diseases, from cancer, diabetes and cystic fibrosis to blindness, obesity and alcoholism. Indeed, creating transgenic mice has become something of a cottage industry, Mario Capecchi told his audience in Stockholm during his Nobel lecture. That is largely thanks to his own obstinacy when, in 1980, he approached the National Institutes of Health for funding to develop his new technology and was told to forget it; his chances of success in applying it to mammalian cells were vanishingly small and he should give up. Convinced he was on to a good thing, Capecchi took no notice and soon his whole lab was working on the project. When it became clear in 1984 that their experiments with mammal cells were working, he applied again to the same department at the NIH for funds. This time he was successful, and the NIH had the grace to say, in their letter of approval, 'We are glad that you didn't follow our advice.'

MARIO CAPECCHI AND THE 'KNOCK-OUT' MOUSE

As far as p53 research is concerned, one of the most valuable transgenic models has been the so-called 'knock-out' mouse, in which a specific gene is deleted from the mouse's DNA to see how the animal functions without it. While the three Nobel winners are collectively known as 'the fathers of transgenic mice',* the knock-outs, which are created using a modified version of the gene-targeting technology, are the brainchild of Capecchi, a man whose journey towards the pinnacles of science no novelist could

* In fact, the first ever transgenic mouse was created in 1982 by Richard Palmiter and Ralph Brinster working at the Universities of Washington and Pennsylvania respectively. But genetic modification was made a great deal easier and more precise by the technology that won Capecchi and Smithies their Nobel Prize.

have made up convincingly. Capecchi lived rough on the streets of war-torn Italy for five years from the age of four, and didn't go to school until he was nine.

He was born in Verona in 1937 – a time when Fascism, Nazism and Communism were raging throughout the country, he wrote in his autobiographical sketch for the Nobel Committee. 'My mother, Lucy Ramberg, was a poet; my father, Luciano Capecchi, an officer in the Italian Air Force. They had a passionate affair, and my mother wisely chose not to marry him.'

Capecchi's mother studied at the Sorbonne in Paris, where she became politically active, joining the Bohemians, a group of poets who openly opposed Fascism. She returned to Italy in 1937, giving birth to Mario in October of that year and settling eventually with him in a chalet in the Alpine Tyrol. Fearing that her activism would mark her out, she began saving money to enable her neighbours, an Italian peasant farming family, to take care of her child if she was taken away.

'In the spring of 1941, German officers came to our chalet and arrested my mother. This is one of my earliest memories,' writes Capecchi. 'My mother had taught me to speak both Italian and German and I was quite aware of what was happening. I sensed that I would not see my mother again for many years, if ever.' Aged three and a half he moved in with the family next door and joined in with the simple life of the farm. 'In the late fall, the grapes were harvested by hand and put into enormous wooden vats. The children, including me, stripped, jumped into the vats and mashed the grapes with our feet. We became squealing masses of purple energy. I still remember the pungent odour and taste of the fresh grapes.'

Capecchi remembers also the day when American warplanes flew low over the fields, 'senselessly' machine-gunning the peasants. But he does not remember how or why the money for his support ran out, only that at the age of four and a half he had to leave the farm. 'I set off on my own,' he writes. 'I headed south, sometimes living in the streets, sometimes joining gangs of other homeless children, sometimes living in orphanages, and most of the time being hungry. My recollections of those four years are vivid . . . Some of them are brutal beyond description, others more palatable.'

Capecchi's mother survived the German prison camp and set out to look for her son, finding him in October 1946 in a grim hospital in Reggio Emilia where he was being treated for malnutrition and typhoid. The two travelled together to the US, to join Lucy's younger brother Edward Ramberg in Pennsylvania, where he was living with his wife Sarah in a commune. Sarah taught the young Mario to read and write, and the boy now went to school for the first time. His Uncle Edward was a physicist renowned for his part in developing the first electron microscopes, and Capecchi himself took physics and maths when he went on to college. He found his studies intellectually satisfying, but rooted too much in the past. He was looking for the challenge of the new and 'a science in which the individual investigator had a more intimate, hands-on involvement with the experiments'.

He found both during a three-month work-study programme at MIT. It was the late 1950s. 'There I encountered molecular biology as the field was being born,' he writes. 'This was a new breed of science and scientist. Everything was new. There were no limitations. Enthusiasm permeated this field. Devotees from physics, chemistry, genetics and biology joined its ranks. The common premises were that the most complex biological phenomena could, with persistence, be understood in molecular terms and that biological phenomena observed in simple organisms, such as viruses and bacteria, were mirrored in more complex ones.'

Capecchi was hooked. He applied to Harvard, 'the perceived Mecca of molecular biology', for his graduate studies and was taken on by James Watson – discoverer with

Francis Crick of the structure of DNA in 1953 – who had told him when he asked for advice that he would be 'fucking crazy to go anywhere else'. Capecchi observed, 'The simplicity of the message was very persuasive.' He was in his element at Harvard and flourished under the 'merciless' but fair and extremely supportive mentorship of Watson. 'Doing science in Jim's laboratory was exhilarating,' he writes. 'As an individual, he personified molecular biology and, as his students, we were its eager practitioners. His bravado encouraged self-confidence in those around him . . . He taught us not to bother with small questions, for such pursuits were likely to produce small answers . . . Once you made it through Jim's laboratory, the rest of the world seemed a piece of cake. It was excellent training.'

After receiving his PhD, Capecchi spent another six years at Harvard before moving for the big skies and rugged open spaces of Utah, to join a new young Department of Molecular Biology being established at the university by scientists he admired and whose vision he shared. He has remained at Utah ever since, and it was here that the knock-out mouse was created in 1989. It was just such a mouse that Scott Lowe and Tyler Jacks used to investigate p53's role in apoptosis.

WHAT HAPPENS IF WE 'KNOCK OUT' p53?

'The University of Wisconsin is an agricultural school and our animal model was the pig. So in addition to learning how lipoproteins* interact with their receptors and how failures in that lead to high cholesterol, I also got experience with animal models – and particularly big ones.' Scott Lowe, a fit-looking, ruggedly built man in his late forties, smiled as he described, in a deep, melodious Midwestern

---

* A lipoprotein is a combination of a fat and protein molecule. The protein helps to transport fat to where it is needed in the body.

accent, his entry into molecular biology. Lowe didn't much like science at high school and imagined he would become a lawyer. But he had taken the opportunity, while studying biochemistry and genetics among his courses as an undergraduate, to spend time in a lab and discovered he loved it: asking questions and dreaming up ways to test ideas were fun. Research, he realised, was what he wanted to do, and after graduating from Wisconsin he managed to get into MIT, the hothouse of bright minds and exciting science, where Mario Capecchi had discovered his calling some 30 years earlier.

At MIT Lowe met Tyler Jacks, a young researcher who had picked up Capecchi and colleagues' new technology with enthusiasm and was busy creating transgenic mice of all kinds to investigate cancer-related genes. Jacks had made some knock-out mice in which various tumour suppressors had been deleted and he was asking the simple and obvious question: do the animals get cancer? He had a mouse model with p53 knocked out, but he had been beaten to it in his experiments by another scientist who had been investigating the same question, so his p53 knock-out mice were sitting around with not much to do. Jacks was happy to let Lowe suggest alternative experiments with them.

Lowe was already fascinated by apoptosis. He had done some work with cell cultures, watching it happen, to his great surprise, in response to oncogene activity, and he was not sure what role, if any, p53 was playing. Perhaps the knock-out mice could answer this question. He had also seen Oren's paper in *Nature* about his temperature-sensitive mutants and the experiments he had done with leukaemia cells which killed themselves when the thermostat was set too low and his mutant p53 morphed into wild type. 'Moshe Oren's experiment was very exciting, but it over-expressed the p53 gene, and one always has to worry that that might be artefactual – you don't know if it does what it appears to do in real life,' Lowe told me when I visited him at Memorial

Sloan Kettering Hospital, where he has a lab on the eleventh floor looking out over the dramatic roofscape of New York City. Over-expressing the gene is like using a sledgehammer, I suggested. 'Yeah, that's right. Cells are sick, they die, right? So it wasn't clear. Particularly since the view was still very strongly that p53 was a checkpoint gene.'

To test whether p53 induces apoptosis in real life Lowe decided to concentrate on the thymus gland, a particularly sensitive organ in which programmed cell death was already known to play a role. The thymus gland produces the thymocytes, important cells of the immune system that give rise to the T cells. Autoimmune diseases occur when T cells start to attack the body's own cells rather than fending off invading organisms. Scientists had discovered that to prevent this happening, thymocytes that might cause problems are weeded out naturally by apoptosis as part of the thymus gland's regular cycle of production and quality control. Barbara Osborne, an immunologist on sabbatical at MIT, had suggested this might be an ideal system for Lowe and Jacks to study.

There are many stimuli known to drive thymocytes to commit suicide, explained Lowe. 'So we decided to line up all of the treatments that were known to trigger this response and then compare the normal mice with the p53-deleted ones, to see if any were defective when p53 was gone. We treated them a lot of ways and mostly the cells died normally, whether p53 was there or not. But the one treatment that was different was radiation, which is known to damage DNA. You could take a machine that would produce gamma rays – the same kind of rays you'd use to treat patients in radiotherapy – and if you irradiated the p53 knock-out cells, they didn't die nearly as effectively as the others. So that was sort of genetic proof that p53 was critical for an apoptotic programme – not as a general principle for all cells, but a very specialised subset: ones which had been irradiated.

'And so it fit the model in a way. That DNA damage

could activate p53 was known, but in all the other cell types tested so far it led to a checkpoint arrest. Here the cells *died*.' This was decisive evidence that p53 was involved in apoptosis in real life, and it was an enormously important finding. Lowe and Jacks – whose p53 knock-out mice had also helped Kastan confirm his hypothesis that p53 leaps into action when DNA is damaged – were riding high. 'We knew instantaneously that this was a huge result, and that it was going to sail into a very prestigious journal,' said Lowe.

Then came a crushing blow. On the far side of the Atlantic, Andrew Wyllie had teamed up with Alan Clarke, another transgenic mouse man, at Edinburgh University; they had been working on exactly the same experiment with thymocytes and come up with exactly the same results. In the race to publish first the two teams came neck and neck; both had submitted their papers to *Nature* and, in a move that took Lowe by surprise, both papers appeared together in the same edition of the journal in April 1993. 'There was a nice summary by David Lane that highlighted how important this result was . . .' concluded Lowe gamely, unable to hide a note of keen disappointment even a couple of decades later.

As Wyllie tells it, the discovery of p53's role in apoptosis following radiation of the thymus was for him also one of the most thrilling moments in his scientific career. 'And we so nearly missed that one!' he mused when I spoke to him at the meeting in Sheffield. Wyllie's inclination was to bombard the p53 knock-out mouse with steroids as a killing stimulus for the thymocytes, because this was closest to a natural scenario. But, significantly, steroids don't cause DNA damage. 'So we used the steroid and got nothing. There was no difference in effect between having p53 and not having p53. The cells died on schedule the same as the controls.' So Wyllie and Clarke decided, belatedly, to try radiation on their knock-out mice thymus cells as well, since DNA damage was the hot topic at the time. 'The effect was entirely different,' he commented. 'In the presence of radiation and p53 you got beautiful, reproducible cell death, which was apoptosis. And if you did the same experiment with radiation and without p53 the cells did not die . . . That was a golden moment, absolutely! That paper was written over a weekend,' said Wyllie with a grin.

ENGINEERED MOUSE RAISES TRICKY QUESTIONS

Fast-forward some dozen years to the first decade of the 21st century and the lab of Gerard Evan, whom we first met at the start of this book, marvelling at the extreme rarity of cancer in the multitudinous cell population. Evan is something of a maverick. From time to time he is prone to dropping bombshells that shake the foundations of prevailing opinion, and he did so in 2005–6. Recognising the vital importance of understanding what each component of the cell does in the greater scheme of things, he is an enthusiastic mouse man. 'The cells in a tumour are all interacting with one another. It's not as if they're all doing their own thing and just decking it out,' he explains. 'They're all talking to one another, reacting and interacting with the normal cells in the body, and educating and instructing them. And that's just by virtue of the fact that as a multi-cellular organism the way we hold together is that our cells talk to each other all the time.'

Evan adapted the gene-targeting technology to make an even more sophisticated version of the p53 knock-out mouse. He replaced the animal's natural p53 with a doctored version of the gene that he could toggle at will between two states – from inactive to functional and back again – by giving and withdrawing a specific drug (a hormone) that controls the 'silencer' on the gene.

p53 as the 'guardian of the genome' that leaps into action in response to DNA damage was well established. But Evan's experiments had led him to question whether its function as a tumour suppressor was as straightforward as that paradigm suggested. He wanted to test it – and in so doing he proposed a heresy. For his experiment he needed a cancer known to be induced by DNA damage, and he chose leukaemia because he knew from experience that if p53 is not functional, mice will develop this type of cancer very quickly after radiation. He knew too that, conversely, mice will be protected from leukaemia even after radiation so long as p53 is present and functioning properly. That much was clear.

But Evan and his team wanted the answer to a very simple question: 'Do you need p53 around at the moment the DNA is sustaining the damage that causes the cancer in order to protect against cancer?' Or to put it another way, at what stage during the gradual development of a tumour does p53's activity become critical to protecting us from runaway disease? Teasing out the answer, however, was not quite so simple. It involved looking at two different scenarios with their engineered mice – one in which p53 was active and functioning during radiation treatment, and the other in which the gene was silenced during treatment.

The experiments for scenario number one brought no surprises at first; the scientists saw what they expected: the animals got very sick, there was 'mega-death' in their lymphoid organs, their bone marrow, their gut – all sites of fast-dividing cells. 'This was always thought to be the price you pay for p53 getting rid of the damaged cells, right?' comments Evan. 'So you get all of this: animals get sick from the mass dying of cells, but they recover.' However, once p53 had done its job in response to the DNA damage caused by the radiation, the researchers 'switched the gene off' again, and here they did get a surprise. 'Blow me down, the incidence of cancer was like you never had p53 there at all! So none of that pain had any gain in terms of tumour suppression – it was, like, *irrelevant*.'

Then Evan and his team did the opposite experiment: they kept p53 silenced and out of the picture when they irradiated the mice, and looked to see how they reacted. The scientists were not surprised initially to find that the animals didn't get sick, because, with p53 inactivated, there was no mass suicide of cells. But what happened next stopped them in their tracks. When, after giving the mice time to repair the DNA damage to their cells, they switched p53 back on again, they found that the mice didn't get sick – again, because there was no mass suicide of repaired cells – but to their great surprise the animals didn't get cancer either. In other words, p53 was able to keep cancer in check, even when it was introduced well after the damage to the DNA had been done.

So what did they make of this extraordinary picture? What this means, explains Evan, is that if you don't have p53 around when the DNA damage is occurring, most cells will be repaired. In real life this is a rather hit-and-miss affair, a patch-up process that leads to the kind of 'mistakes' that are the driving force behind evolution. But if you then restore p53 after this repair process has had time to work, the gene will *only* be activated in those cells that have sustained mistakes, or mutations, that make them dangerous (for example, ones that activate oncogenes) and therefore send out alarm signals to abort. 'This basically means that you can separate out the DNA damage response from the tumour suppressor response,' he comments.

This has enormous implications for treatment of cancer, because it implies that we could devise ways to prevent most of the dreadful side effects of chemo- and radiotherapy – the hair loss, nausea, exhaustion, immune suppression – that are the direct consequence of hitting all the body's fast-dividing cells, and clear out just those cells that go on to become cancerous and that therefore continue to send out alarm signals that activate p53. But how did Evan's colleagues react to his findings and his theories?

'You know ideas like this take a long time to percolate through. I mean getting it published . . . I remember one of the reviewers just said, "I refuse to accept that the DNA response is not the major tumour suppressor pathway." But this is not a faith-based thing; we're not a religion! These are the data. And I wasn't saying, "You're all wrong." I was saying, "These are the data. This is our explanation. This is our hypothesis." The whole point about publishing in the literature is that you publish the data; you publish the hypothesis so that it can be tested by the community.

'That experiment was a very intriguing experiment, and I think a very informative one. And I think the conclusions of it still stand. But the point is that cancers arise in many different tissues and many different ways, and the issue for me is much less about are these data wrong or are these data right than about getting to understand which set of rules apply in which case.'

A FINE BALANCE BETWEEN LIFE AND DEATH

Evan had met even stronger resistance to his ideas some years earlier when his research into oncogenes suggested to him that all our cells contain both growth and suicide programmes that are in constant, hair-trigger competition. Which course of action a cell takes is essentially controlled by its environment and the signals it receives from its neighbours: is it in the right place and at the right time? Is it behaving normally? If so, it will receive 'stay alive' signals; if not, it will be instructed to abort. This is an inbuilt defence mechanism and one of the reasons cancer is so rare, believes Evan.

Since this story takes us back among the Petri dishes in the lab, it may seem like a diversion from the topic in hand, animal models. But besides describing another pivotal moment in cancer research, it helps show why it is so important that experiments are performed in living organisms as well as in cell and tissue cultures. It begins in the late 1980s, when Evan, newly recruited to the ICRF in London, was doing some experiments with the powerful oncogene Myc, looking at how it drives cell proliferation. 'I made this bizarre observation that when you expressed Myc at high levels in cells they did indeed proliferate – but when you looked a couple of days later, there were fewer cells than before,' he explains. 'I'm a great believer in personal observation – observing things, you ask questions of what you can actually *see*. So we took these cells and we put them under a microscope and we used time-lapse video to take one frame every three minutes. Then you speed it up and watch what happens over three or four days in just two or three minutes. And there we saw this amazing phenomenon . . . the cells were replicating, but also they were dying by apoptosis.'

This was exciting, but it didn't make sense. Then a thought struck Evan that drew on his early training in immunology, where a common theory was that the immune system plays a part in protecting us from cancer by eliminating rogue cells that it recognises as foreign. 'I thought what if, instead of the immune system acting as a police service to find aberrantly proliferating cells, there is, hard-wired into the very warp and weft of how cells proliferate, an abort programme? Every time you pick up the machinery to proliferate, you also pick up the machinery to kill yourself?'

If that were the case, he reasoned, there must be something that tells the cells whether to live or die, and here he found a clue in the growth medium he was using for his experiments. Most of the time, he used serum – the colourless liquid the body produces at the site of a wound that makes it 'weep' – because serum contains substances that promote clotting of blood, and survival, growth and proliferation of cells to help in the recovery and regeneration of injured tissue. Myc was killing cells when Evan removed the serum with all its life-enhancing properties and put the cells in a medium that was more like what they would find under normal conditions in the body.

'Myc turned out to be, I think, the first example of what

we now know as a generic feature of how growth control is orchestrated within our cells – which is that everything that makes a cell proliferate (and is potentially therefore a cancer risk if it gets mutated and stuck in the "on" position) comes with something that also suppresses the expansion and growth of those cells.'

Similar experiments with other oncogenes showed that they too shut down growth programmes one way or another after a short spurt of proliferation. Ras, for example, does it by permanently arresting, but not killing, the cell – putting it in a state known as 'replicative senescence', where it stops dividing but stays alive and active. But this raised a number of further questions. Oncogenes like Myc and Ras, when not mutated, have regular work to do in promoting growth in cells, but if they also serve to shut down or kill cells after a while, how is new tissue ever produced? 'The answer seems to be that if a cell switches on Myc in response to a growth signal and starts to replicate, if that cell is in the right place in the body, and it stays in its little niche and doesn't spill out like a cancer, then it will get all the goodies that tell it not to commit suicide, okay?

'So cell replication is an obligatorily social enterprise. Cells are not autonomous. By taking them out and putting them in a bottle and adding all the things that would stop them dying, we just completely ignored this fundamental piece of biology. It had always been ignored! Now, the notion that things that drive cell growth also drive cell death and growth arrest is, I think, completely embedded in the understanding of molecular biology; it's just generally accepted that this is how things work. But at the time, people literally walked out of my talks!'

In fact, Scott Lowe, then doing his PhD at MIT, and his supervisor Earl Ruley, had observed the same extraordinary phenomenon – oncogenes killing cells or condemning them to replicative senescence. And they too had had a tough time getting people to listen. 'If you'd walk down the hall at MIT

Cancer Center and say, "I have an oncogene and it kills cells", they'd think you were crazy. Because that's not what oncogenes do; they make cells grow better,' laughs Lowe.

The insights he and Evan gained in this work also helped to explain a long-standing mystery: *why* oncogenes are only able to generate tumours in co-operation with one another. Evan believes that when, for example, you put Myc and Ras together, Myc overcomes the replicative senescence programme of Ras, and Ras overcomes the apoptosis programme of Myc. Thus singly, the growth spurt fuelled by either oncogene soon fizzles out; together, all hell breaks loose. In time, he and Lowe would discover that the effects they had both witnessed independently and wondered about – death among their oncogene-driven cells – were caused by the oncogenes switching on tumour suppressors, frequently p53.

The multiple experiments with mouse models – knocking out p53 altogether, or else toggling the gene back and forth between active and passive – made it very clear that this is an extremely powerful protein. As an arbiter of life-and-death decisions within our cells it must be under strong control. So how does it work?

CHAPTER THIRTEEN

The Guardian's Gatekeeper

In which we learn: a) that the enormously powerful p53, which can arrest or kill cells, is kept on a tight leash by a protein called Mdm2, which sticks to the p53 protein in cells and marks it up for degradation; and b) that Mdm2 releases p53 from its deadly embrace only when the tumour suppressor is needed to respond to stress signals.

***

The reason that cancer research is such a compelling area to be in is that, in order to understand how things go wrong in cancer, you first have to understand how things go right almost all the rest of the time.

Gerard Evan

'Guardian of the genome', the epithet David Lane gave to p53 in 1992, caught the popular imagination and has helped give this extraordinary gene with the eminently forgettable name a public profile in the media. But in the 20 years since that phrase was coined, the list of stresses to which p53 responds has expanded way beyond simple damage to the DNA. Arnie Levine suggests it might be more appropriate to think of the gene as a 'fidelity factor' – something that ensures faithful copying of the DNA during cell division. We now know that the gene responds to heat shock and cold shock (when a cell is subjected to temperatures above or below the ideal body temperature), to lack of oxygen or glucose, to certain poisons, to natural ageing and to oncogene activity – all things that threaten the fidelity of the DNA, without actually breaking it, as the cells divide.

'Ever since its evolution in invertebrates, p53 has been a fidelity factor,' says Levine. Starve a worm of glucose and it won't produce eggs; radiate a fruit fly and it won't produce sperm or eggs until progenitor cells with healthy genomes are restored. With the evolution of the vertebrates – organisms with much more complex bodies, including us – the principle of protecting the fidelity of the germ cells, the sperm and eggs that give rise to offspring, was also applied to all other cells of the body for the first time. 'And that's where p53 comes in,' says Levine. 'It responds to stress, and it kills. It enforces fidelity by death! So it has a very interesting evolutionary history.'

Very recently scientists have discovered another fascinating aspect of p53's role in fidelity assurance. Imagine for a minute what would happen if, in the normal course of events, our biological clocks could go backwards in time; if our mature cells could revert to their original undifferentiated state as stem cells, complete with the potential to develop afresh into something new. It's a nightmare scenario in which your liver cells might morph spontaneously into bone cells, gut into teeth, blood into kidney, and no bodies would be stable. It is p53's job to ensure that such de-differentiation doesn't happen; that biological time moves inexorably forward and that our bodily development cannot unravel (except, that is, in the deranged environment of cancer). Scientists creating what are known as 'induced pluripotent stem cells' (IPSs) – stem cells with the potential to become any kind of specialised cells that have been engineered in the lab from already differentiated body cells – are frustrating a fundamental law of nature, and they must overcome p53's defences to do so.

Many people have been involved in uncovering the mechanism controlling this powerful gene, and once again Moshe Oren's temperature-sensitive mutants provided vital insights. This was the early 1990s, before the creation of knock-out mice had become a cottage industry. Oren and his team were using his temperature-sensitive mutant p53 as a tool in cell cultures to ask a simple question: what is different in cells with active p53 compared with those in which the gene is inactive? They soon observed that a protein appeared hitched to the p53 protein whenever the tumour suppressor was behaving like wild-type p53, at 33°C (91°F), but never at the higher temperature, when it behaved like a mutant.

Levine had observed the same thing in a different set of experiments designed to explore the workings of wild-type p53, and he had identified the hitchhiking protein as Mdm2, already known to cancer researchers as a possible oncogene. Levine had also discovered that by attaching itself to p53, Mdm2 restrained the tumour suppressor – more like a policeman handcuffed to a criminal suspect than a hitchhiker. Tinkering with the temperature in their Petri dishes, Oren's team discovered that Mdm2 protein appeared only in the cells with normal p53, and was absent altogether from the mutant cells. What did this mean?

Oren and company soon realised they had in their hands a crucial piece of the jigsaw that would reveal the control mechanism of p53. Timing was all. 'This was 1993, and a year earlier we wouldn't have known what to do except to say that this was very interesting,' he commented. But just the year before, Carol Prives and Bert Vogelstein had shown that p53 is a 'transcription factor' whose job is to switch other genes on and off; Vogelstein had identified the first of its 'downstream' targets and it seemed reasonable to suggest that Mdm2 was another – that it depended on wild-type p53 to switch it on. Subsequent experiments proved their hypothesis right, as Oren explained: 'Knowing what Mdm2 does to p53 in the way of acting as an inhibitor, which was Arnie's discovery, and seeing that p53 itself is activating Mdm2, we discovered this "feedback loop" – which luckily turned out to be not just interesting but extremely important, because this is the main way in which p53 is regulated in the cells. This Mdm2 feedback loop is perhaps the heart of the p53 regulatory network.'

In fact, the feedback loop – in which the inhibitory

protein is switched on by the protein it then inhibits – did not quite make sense yet. There were still two vital pieces of the puzzle missing, and Oren's lab was again at the forefront of the search. They had started to use Mdm2's ability to restrain p53 as a tool much like the temperature-sensitive mutant: by adding or removing the gatekeeper from cells in which p53 was present, they could switch the protein between active and inactive states. 'We were using it routinely to try to get insights into what p53 does, which is what *doesn't* happen when Mdm2 inhibits it,' explained Oren. However, the researchers soon started to notice something puzzling: that whenever they put the two proteins together, instead of Mdm2 simply attaching itself to p53 and stopping it in its tracks, as they had expected, the p53 protein seemed to disappear altogether. Initially the scientists didn't trust their experiments, so they repeated them with slight modifications and extra care. But when they came up with the same results again and again, they knew they were real: Mdm2 was destroying p53. And it was doing so, they discovered later, by delivering the 'kiss of death' – attaching a little chemical tag to the p53 protein that marked it out for collection and degradation by the cell's recycling machine, the proteasome.

So, the picture Oren's team had built up thus far showed that p53 switches on Mdm2, which in turn marks p53 up for degradation in an endless cycle, lasting around 5–20 minutes, that keeps p53 protein in our cells at almost undetectable levels most of the time. (This incidentally helped to explain why there were such high levels of the protein in cells in those early experiments with the mutant clones: mutation typically severs the bond between p53 and its controller Mdm2). But how then does the tumour suppressor escape its gatekeeper in order to protect us from cancer in the normal course of events? This was the final missing jigsaw piece, and it was discovered tucked away in a corner of Carol Prives' lab in Columbia, where a 'fantastically talented' young postdoc called Sheau-Yann Shieh from Taiwan was interested in a process called phosphorylation.

Phosphorylation is one of the most important mechanisms by which proteins are activated and silenced in cells, and it works by attaching small phosphate molecules, as 'tags', to some of the amino acids that make up the protein. Shieh was focusing on p53 and asking the question: does the protein get phosphorylated? If so, how does this affect its function? She found that p53 does indeed become phosphorylated; it changes shape and, incidentally – for she was not paying special attention to the Mdm2 story, which was still evolving – that this weakens the bond between the p53 and its minder.

'But we were missing a critical link,' explained Prives. The pressing questions that remained were: what is the trigger for p53 to become phosphorylated? Does this happen in real life? (So far they had seen it only in cell cultures in the lab.) And is it related to DNA damage and other stressful events in the cells? With the help of some fancy new tools for singling out phosphorylated proteins, they found the answer to all three questions. They discovered that phosphorylation of p53 does happen in real life. It is indeed related to DNA damage and the stress response. And it is the mechanism by which ATM – the gene that makes patients with ataxia telangiectasia so acutely sensitive to radiation – signals distress and triggers the response from p53.

This piece of the jigsaw completed the picture of the p53/Mdm2 feedback loop and suggested how it might work. In essence, this is how it looked: in the normal course of events, p53 protein is being made in our cells all the time so that it can give a hair-trigger response to danger signals; and it is being cleared away almost as fast as it is made by Mdm2 lest it lead to unnecessary death of cells. But danger or stress activates genes such as ATM, which phosphorylate p53, preventing Mdm2 from getting a proper hold and allowing the p53 protein to accumulate in the cells and perform its job of orchestrating the response – arresting the cell temporarily and sending in the repair team; condemning it to permanent arrest or senescence; or forcing it to commit suicide.

Another researcher, Gigi Lozano, working with mouse models at MD Anderson Cancer Center in Houston, confirmed just how important each protein is to the normal functioning of the other in real life when she created mice with Mdm2 knocked out, and discovered that this was biologically lethal: with no holds barred, p53 caused massive apoptosis. But when Lozano created mice with both p53 and Mdm2 knocked out, they could survive until, eventually, they developed cancer.

Exactly how the stress-response mechanism is turned off again when it is no longer needed, and how the tight regulation of p53 is restored, no one is completely sure as yet. But it could be that when the stress signals stop coming, any newly produced p53 will not be protected from degradation by phosphorylation, and the protein will resume its normal dance of death with its minder, Mdm2. This remains an important question since the feedback loop offers tantalising opportunities for tinkering with the regulation of p53 to create new cancer treatments. It's a topic I will return to in a later chapter.

CHAPTER FOURTEEN

The Smoking Gun

In which we discover that the component in tobacco smoke that causes lung cancer, benzo(a)pyrene, does so by sticking to DNA and damaging the p53 gene, leaving a mutation that is as unique and incriminating as a fingerprint.

***

When a risk factor for a disease becomes so highly prevalent in a population, it paradoxically begins to disappear into the white noise of the background . . . If nearly all men smoked, and only some of them developed cancer, then how might one tease apart the statistical link between one and the other?

Siddhartha Mukherjee

In the mid-1990s, p53 emerged briefly from the rarefied environment of academia to play a dramatic role in the battle between public-health authorities and the tobacco industry by decisively nailing the connection between smoking and lung cancer. The connection itself was not news. Nearly half a century earlier, in 1950, the Oxford-based epidemiologist Richard Doll – a maverick and somewhat controversial figure among the medical establishment because of his membership of the Communist Party – had drawn attention to the association between cigarettes and lung cancer in a paper for the influential *British Medical Journal*. In it, Doll and his collaborator, Austin Bradford Hill, reported their findings from a research project aimed at figuring out what was causing the sudden explosive rise in lung-cancer cases in the UK.

In the quarter-century from 1922 to 1947, deaths from lung cancer in England and Wales rose from 612 to 9,287 per annum – 'one of the most striking changes in the pattern of mortality recorded by the Registrar-General', Doll said in his report. Tobacco had been a rare source of comfort to men fighting in the trenches and on the high seas of World War I and cigarettes were included among a soldier's or sailor's rations in both the British and American armed forces. Indeed, General John Pershing, leader of the American Expeditionary Forces in World War I, is reported as saying, 'You ask me what we need to win this war. I answer tobacco as much as bullets . . . We must have thousands of tons without delay.'

Thus a generation of young men returned to civilian life addicted to tobacco, entrenching a habit that had been growing since the invention in the late 1880s of a machine for rolling cigarettes that enabled mass production. By the time Doll and Hill were doing their studies in the late 1940s, approximately 80 per cent of British men were regular smokers and the death rate from lung cancer in England and Wales had increased more than sixfold among adult men and approximately threefold among women in less than 20 years.

The two scientists did not, however, suspect tobacco of being the cause of the premature deaths. Their initial hunch was that it was atmospheric pollution – dust from the new road-building material, tarmac; exhaust fumes from cars, power stations and the coal fires burning in everyone's front rooms. But the picture of smoking habits that emerged from the questionnaires conducted with 1,732 cancer patients and 743 controls in 20 London hospitals pointed compellingly to tobacco exposure as the culprit. Doll and Hill's *BMJ* paper concludes with the statement that 'above the age of 45, the risk of developing [lung cancer] increases in simple proportion with the amount smoked, and that it may be approximately 50 times as great among those who smoke 25 or more cigarettes a day as among non-smokers'.

Though he had no idea what the carcinogenic substance in tobacco could be, Doll himself gave up smoking on the strength of the epidemiology. As the evidence of its deadly effects accumulated, governments in many countries adopted measures aimed at curbing the habit. But as long as the link between smoking and lung cancer remained circumstantial, without a physical mechanism to back it up, cigarette companies had ample wriggle room to fight the public-health case against them. They continued to promote their product aggressively, particularly in developing countries with potentially massive markets and poor controls. Even when scientists, following on from Doll and Hill's work, showed that painting tar – the sticky black residue of tobacco that coats smokers' lungs – on to the skin of laboratory mice causes tumours to grow, Big Tobacco was wont to scoff that the experiments were irrelevant: these were mice, not people.

In fact, the first person to do such an experiment was Angel Honorio Roffo, an Argentinian oncologist who flagged up the link between smoking and cancer nearly two decades before Doll and Hill's paper in the *BMJ*. By painting rabbits' ears and the skin of mice repeatedly with either nicotine or tobacco tar, Roffo identified the latter as being the carcinogenic substance – nicotine alone had no effect on the animals' skin no matter how long he left it. But the results of his extensive research into the disease that afflicted his patients went almost unnoticed by anyone except the tobacco industry. This was at least partly because his most important scientific papers were published only in German and partly because he was way ahead of his time in his understanding of cancer biology. So far ahead, in fact, that a New York physician who had heard of Roffo's studies and written to the American Tobacco Company (AT) in May 1939 to ask if his findings were valid was able to be reassured by the cigarette manufacturer's own research director. Hiram Hanmer replied to the doctor that AT had been following Roffo's work for some time and felt that it had left the literature on tobacco 'in a very beclouded condition'. He assured his correspondent that 'the use of tobacco is not remotely associated with the incidence of cancer'.

Rubbishing the science has always been the tobacco industry's *modus operandi*, but it became infinitely more difficult – and finally impossible – once the molecular biologists were on the case. Curt Harris is a bear of a man with a shaggy grey beard and a deep bass voice as rich as a pint of brown ale; he heads the Laboratory for Human Carcinogenesis at the National Cancer Institute in Bethesda. He is also a pioneer of the relatively young field of molecular epidemiology, the science that traced the origins of the AIDS pandemic back to chimpanzees and green monkeys in the forests of Africa by following the genetic footprint of the virus that became HIV. It's the science that uncovers the sources and assesses the virulence of regular flu outbreaks; and it's a science used widely today to try to pin down the multiple causes of cancer.

In the 1980s, Harris's lab was involved in studying the activity of carcinogenic substances, including components of tobacco tar, when they get into cells. They found that a number of these substances stick themselves firmly to the DNA, and Harris knew from the literature and from his own studies that this would lead to mutations in the genes, the first step along the road to cancer. He became especially fascinated by Bert Vogelstein's work with p53, and the two men decided to collaborate on a research project to assess how commonly p53 might be mutated in cancer. This was 1989, just after Suzy Baker in Vogelstein's lab had made her sensational discovery that p53 was a tumour suppressor, not an oncogene. Harris and Vogelstein found p53 mutations in many of the common tumour types they saw in their clinics, including breast, lung, brain and colon. They found too that the mutations showed a pattern and were clustered in four particular positions along the length of the gene, which they dubbed 'hot spots'.

The two researchers published their findings in *Nature* in

1989. Soon afterwards Harris, who remained intrigued by the pattern he and Vogelstein had revealed, decided to collect systematically the information on the different mutations mentioned in the steady stream of papers that appeared on p53. In 1990, in collaboration with Monica Hollstein, then working at the World Health Organization's International Agency for Research on Cancer (IARC) in Lyon, France, he formalised his collection into the p53 database, a rare resource for scientists and clinicians that can provide clues to the identity of carcinogens in the environment, as well as to the likely course of a patient's disease or the best options for treating certain tumours.

In 1994, Hollstein left IARC and her place at the helm of the database was taken by Pierre Hainaut, who expanded it, increased the level of detail recorded for each mutation and managed it until mid-2012. Today the database contains information on tens of thousands of mutations, together with the tumours in which they occur and as much information as possible about the lifestyles and personal characteristics of cancer patients, including their response to treatment and the final clinical outcome if available.

'We started the database because we had this idea that there was going to be a relationship between environmental causes of cancer and the p53 mutation spectrum,' Harris told me. And indeed to the disease detectives – the molecular epidemiologists – the database has proved invaluable. Its role in the tobacco story is one of its most notable successes.

TOBACCO'S FINGERPRINT FOUND ON p53

In the early 1990s, Gerd Pfeifer, who heads a lab at the City of Hope medical centre in Duarte, California, was exploring DNA damage in relation to cancer, looking to see if the causative agents left distinctive patterns of damage, or 'fingerprints', that identified them as culprits. His lab had developed a tool that enabled the scientists to home in on individual genes among the thousands along a strand of DNA – a process akin to finding needles in haystacks – and the general buzz surrounding p53 at the time persuaded Pfeifer and his fellow researcher, Mikhail Denissenko, that this gene would be an interesting focus for their studies. They would look at the effects of tobacco smoke on p53.

'The early tobacco work had clearly suggested there is high tumour-causing activity in the tar – the black stuff you can collect on filters when you burn tobacco, and that you can see in the lungs of heavy smokers when you operate on them. It looks gross,' Pfeifer commented when I phoned him in his lab. He and Denissenko knew that the most damaging components of tar are the polycyclic aromatic hydrocarbons, or PAHs, of which one, benzo(a)pyrene (BaP), is particularly noxious. This, they decided, would be an ideal damaging agent to use in their experiments.

Others had already investigated what happens to PAHs when they get inside cells. Their research showed that these substances are not water soluble, so the body has difficulty ridding itself of them. In an effort to transform the compounds into something that can then be excreted, the machinery of the cells creates a dangerously reactive substance that sticks itself to the DNA. In the case of BaP, this transformed substance has a mind-numbing formula represented simply as BPDE, which is considered to be one of the most potent carcinogens yet discovered.

Pfeifer and Denissenko took BPDE and added it to various cell types, including human lung, and then left the cells to their fate. After an hour or two they returned to their lab benches to isolate the DNA and to apply their special technique to identifying the exact positions on the p53 gene that had sustained damage. This, they discovered, was not random. BPDE always attached itself to the DNA next to the guanine base – one of the four basic building blocks of DNA, represented by G in the DNA code – at three very specific 'hot spots' along the gene, at codons 157, 248 and

273 (to recap, a codon is a segment of a gene just three bases long that codes for one of the building blocks of the protein, and the codon's number defines where that building block should go).

Here at last was proof that a defined product of cigarette smoke damages DNA. But the clincher for the case against Big Tobacco came when Pfeifer and Denissenko compared their lab's results with the p53 mutation database, which by 1996 had more than 500 entries for lung cancer drawn from the literature worldwide. The great majority of the mutations described in the database among smokers, but rarely among non-smokers, corresponded precisely with their results: they occurred in the same hot spots targeted by the BPDE and they showed that the building blocks of the gene had been scrambled such that the guanine (G) was replaced by a thymine (T). Crucially, while codons 248 and 273 are mutation hot spots in many types of cancer, the database revealed that codon 157 is found exclusively in lung tumours. In other words the fingerprint of BPDE was all over the p53 database. Pfeifer and Denissenko's paper, published in *Science* in October 1996, concluded, 'Our study thus provides a direct link between a defined cigarette smoke carcinogen and human cancer mutations.'

BIG TOBACCO QUESTIONS THE SCIENCE

This was bad news for Big Tobacco. It meant not just that smoking was implicated in a generalised threat to public health, but that it could be linked to individual people with lung cancer. Tobacco companies were now much more vulnerable to lawsuits from customers looking for compensation for ruined lives and, as ever, they set about trying to refute the evidence. In an address to investors, analysts and journalists soon after the *Science* paper came out, the Chief Executive of British American Tobacco Industries (BAT), Martin Broughton, stated, 'There is still a lack of understanding of the mechanisms of disease attributed to smoking . . . The importance of this *Science* magazine study may lie, not least, in the recognition that there are important missing links in the understanding of causation . . . It may lead to further research . . . into the complex process by which a cell becomes cancerous – a process we and others have spent millions in trying to understand for many years now.'

The R J Reynolds tobacco company was even more blatantly dismissive. A public statement issued by the company said, 'That BaP will cause a mutation has been known for a long time . . . The authors themselves describe these findings as a coincidence. The press release's conclusion that these findings are the key to lung cancer is an overstatement.'

Interestingly, R J Reynolds' statement came out on the eve of publication of Pfeifer and Denissenko's study being published in *Science* – clearly suggesting the company had been tipped off in advance.

The following year, Pierre Hainaut and a colleague at IARC, Tina Hernandez-Boussard, carried out a detailed analysis of the spectrum of p53 mutations in smokers as recorded in their database. Their paper for *Environmental Health Perspectives* came to the same conclusion as Pfeifer and Denissenko's: that these mutations carried the fingerprint of BPDE. 'We thought at the time, that settles it,' said Hainaut as we sat together in the front room of his home in Lyon, with its distant views of the snow-covered Alps, talking about the database's role in revealing the causes of cancer. 'You have experimental data; you know what the mutagenic substance is; you can demonstrate its effect very well in the lab and you can show that the people who are exposed to the same substances in real life get the mutation at exactly the same place.'

Hainaut and Pfeifer believed their two papers made an irrefutable case, and they were taken aback two years later to see first one and then a second paper challenging their results as being 'over-interpretations'. 'People had done some analysis themselves of our database to try to prove that we were wrong – that the connection was not there; that maybe tobacco was *helping* mutations occur, but not actually causing them,' explained Hainaut. 'I was shocked. But also I have to say that we're used to trust within the scientific community, and you expect people to be fair. So the first reaction when you see a paper like this attacking your work is, oh my God, I've missed something very important! I've made a big mistake! In fact when the first of these papers came out my director called me into his office and said, "What's this about? You have three days to bring all your data to me for review, because if you've mishandled or misrepresented an issue like that it's a serious matter." So I was in the hot seat somehow!'

To Hainaut's relief, his interpretation stood up to review and he began to wonder about the author of the second paper, published in *Mutagenesis* – a scientist named Thilo Paschke, working for a Munich-based institute, Analytisch-Biologisches Forschungslabor. At this point in our conversation, Hainaut disappeared into his study and returned with a sheaf of yellowing papers held together in a clear plastic-covered folder – a record of the murky drama he found himself being dragged into. 'I had never heard of this guy, never seen a paper by him and I had no idea about the institute he worked for,' he continued. 'So I tried looking up the institute on the internet and I didn't find any website. All I had was an address in Munich, so I called the German telecommunications people and they told me, "It's the address of the German Association of Tobacco Manufacturers."

'My first reaction was relief. I realised, okay, that's a completely unfair attack on our work and I can probably forget it. All I have to do is demonstrate that it comes from people who have a bias they haven't declared.'

But then the plot began to thicken. As part of the general settlement of lawsuits between the American States and Big Tobacco in 1998, the industry was required to make publicly available all internal documents used during the trial. These were organised systematically and posted on the internet at www.tobaccodocuments.org. In early 2001, Hainaut opened his computer at the website in search of information about the journal *Mutagenesis*, whose behaviour he and Pfeifer had found 'strange' when they raised the issue of Paschke's conflict of interests and asked for space to respond to his critical paper. Their request had been turned down and, intrigued to know if anything suspicious lay behind the rejection, Hainaut typed the name of the journal's founding editor, Jim Parry, into the site's search engine and watched wide-eyed as a series of references popped up on screen.

Here was evidence of research and consultancy contracts between Parry and BAT and Philip Morris running over more than a decade. To the fascinated Hainaut, a letter dated 3rd August 1994, from Parry to his main contact at Philip Morris, the company's chief scientist Ruth Dempsey, about a proposed research project was particularly revealing. In it Parry, then a Professor of Biochemistry at the University of Wales at Swansea, advised Dempsey that 'the overhead figure of 40 per cent I have given in the costs can be overcome if money is given to me as a gift, "as a contribution to my research" without specifying how it should be spent. My colleagues tell me that some companies are increasingly appreciating the financial advantage of giving money in this way compared to contracts which specify components of the project.'

Parry ended his letter, 'Please let me know if your people are interested in supporting the type of work outlined, bearing in mind that we can modify the components in various ways to also fit in with your interests.'

Curt Harris, original founder of the database and by now editor of the journal *Carcinogenesis*, had been taking a personal interest in the controversy. Faced with such clear evidence of a conflict of interest, he blew the whistle on Parry, editor of *Mutagenesis,* with Oxford University Press, which publishes both journals. Although she understood full well the seriousness of the allegation against Parry, Janet Boullin, the editorial director of journals at OUP, pointed out that she had limited scope for action since OUP was simply the publisher, not the owner, of *Mutagenesis*. But she immediately strengthened the rules about disclosure of interests for both contributors and editors of all OUP journals. Parry was not prepared to comply with the new rules and he resigned as editor of *Mutagenesis* soon afterwards. But he remained on the journal's editorial board, whose members were not required to sign competing-interest statements.

Among the documents on the tobacco industry website, Hainaut found evidence, too, that Philip Morris was gathering intelligence on his own institution, IARC, home of the p53 database. However, satisfied for the time being with what he had learnt from his own sleuthing, he handed the evidence he had gathered over to Stanton Glantz, an anti-tobacco campaigner in the US.

SCIENTISTS FIGHT BACK

Stanton Glantz, a heart specialist, is Professor of Medicine at the University of California, San Francisco, director of the Center for Tobacco Control and Education, and a scientist with a long history of standing up to the tobacco industry. Originally, his focus was passive smoking, which kills tens of thousands of non-smokers in America alone every year. But his efforts to have smoking banned in public spaces have seen him hassled constantly by industry supporters. 'After every barrage of personal attacks against me in the *American Smoker's Journal* or the American Smoker's Alliance [set up by Philip Morris and others] or these other publications from these groups, I get a series of hate mail,' he told the Public Broadcasting Service (PBS) in an interview for their programme *Frontline* in 1996. 'I have had hate mail, hate faxes, hate e-mail, hate phone calls. I mean the police department here intercepts some of my mail now, and it's a drag.'

Turning up for work on 12th May, 1994, Glantz found dumped on his desk a large box containing several thousand pages of confidential internal documents from the tobacco company Brown and Williamson, a subsidiary of BAT. There was no clue as to who had sent the parcel, simply the name 'Mr Butts' on the 'return address' label. The documents covered a period from the early 1950s to the mid-1980s, and related to such issues as the addictiveness of nicotine, cancer and the company's public relations and legal strategies. Since these were not directly relevant to his own research, Glantz intended passing the documents on to a colleague working in these areas. But after flicking through them for 20 minutes or so he found himself unable to stop reading. 'The thing that sucked me into them was not their potential political or legal import,' he told PBS. 'It was the documents as history, the documents as science. It was just an unbelievable find. As a professor, it would be like an archaeologist finding a new tomb in Egypt or something . . . I mean it's an amazing, amazing story of what was going on inside these cigarette companies during this crucial period.'

Glantz hung on to the documents and worked through them systematically with a team of reviewers, writing a number of papers for journals before pulling them together, in collaboration with colleagues, into a book, *The Cigarette Papers*, published by the University of California Press in 1998. He remained interested in the tobacco industry and its subversive activities, and a few years later began investigating its involvement with p53, adding further detail to what Hainaut had discovered and writing up the story in an explosive paper for *The Lancet*.

Glantz's paper is littered with the names of scientists

paid by Big Tobacco to carry out research on its behalf, frequently with the explicit aim of discrediting the evidence of a causal link between smoking and cancer, or to write letters to medical journals or popular newspapers for the same purpose. Sifting through the wealth of information the tobacco industry has been obliged to publish, Glantz found evidence that by the early 1990s, BAT had identified p53 as being especially important, with 'more papers published on it than any other topic on cancer'. The company monitored research on the gene and pressed its paid scientists for intelligence about what their colleagues were discovering and for advance copies, if possible, of relevant papers submitted for publication. Furthermore, BAT regarded information about the research organisations it supported as confidential and advised individual scientists that they were 'free to publish their work without further reference' to the company. BAT and others also appear to have anticipated Pfeifer and Denissenko's findings about the effects of tobacco tar on p53, and to have worked on a strategy to discredit them in advance.

The idea to challenge the scientists' interpretation of data from the IARC database seems to have come from Jim Parry, who published the two critical papers in *Mutagenesis*. But though Hainaut was relieved when he first discovered the evidence of skulduggery, he found himself unable to dismiss the attacks out of hand. 'I think, to be fair, the tobacco industry criticism had a point,' he told me. 'Our data were based on putting together bits and pieces of our database from studies which were not aimed at demonstrating what we wanted to demonstrate. We had data on never-smokers, for example, that were scattered around something like 20 different papers, none of which had the scope on its own to demonstrate what we wanted to show; it was just by putting them together that they made the point.'

He and his fellow researchers decided to go back to the drawing board and come up with an analysis that was rigorous and powerful enough to prove their case. 'It took us two years, but we did it and we published the paper in 2005 in *Cancer Research*,' said Hainaut. 'Nobody now can say the connection is not there; it's really watertight. I think the whole story is now behind us, but that's what it took! And maybe in that respect it was a good thing we had this attack, because otherwise we might not have done the paper.'

Finally nailing Big Tobacco more than half a century after the first warnings of the dangers of smoking was a triumph for p53 and public health, but it's not the only one. The mutant p53 database is proving a rich resource for the disease detectives, who are finding a growing list of carcinogens that leave their unique fingerprints on the cancers they cause. Besides tobacco, mouldy peanuts on liver cancer and sunlight on skin are just two of the direct relationships the scientists have been able to work out in forensic detail.

CHAPTER FIFTEEN

Following the Fingerprints

In which we learn that besides lung cancer, other types of cancer, including liver and skin, frequently have mutant p53 that carries the unique fingerprint of the agent that caused the disease.

***

The most exciting phrase to hear in science, the one that heralds the most discoveries, is not 'Eureka!' (I found it!) but 'That's funny . . .'

Isaac Asimov

Liver cancer is the seventh most common cancer worldwide, but in South East Asia and sub-Saharan Africa, where the great majority of cases occur, it kills more people every year than any other tumour type. In these regions Hepatitis B, which is a major risk factor for liver cancer everywhere, is extremely widespread – passed between sexually active adults and from mother to child, much like HIV. And like the AIDS virus, too, it can wreak havoc in a person's body without them being aware of the infection, and become endemic in communities. Hep B generally takes many years to cause liver cancer, but in the high-incidence countries of Asia and Africa people's risk of getting the disease is compounded by exposure also to aflatoxin, a poison produced by the fungus *Aspergillus* that flourishes on peanuts and grains stored in warm, damp conditions without adequate ventilation.

Aflatoxin is a known carcinogen and was one of the chemicals investigated by Curt Harris's lab in the late 1980s and early '90s for its mechanism of action in human cells. As with BaP in tobacco tar, Harris knew that aflatoxin is metabolised and transformed in cells into a substance that sticks to DNA and causes mutations. But it was work he did with colleagues in China, analysing the genetic mutations in liver tumours in Qidong county, on the north side of the Yangtze River opposite Shanghai, that showed the poison at work in the real world and pointed the finger at p53 as being the target for mutation. Rates of liver cancer in the county were exceptionally high; so too was people's exposure to aflatoxin from mouldy grains and beans in their diet, and the researchers were struck by the frequency of an unusual mutation in p53 at codon 249. This resulted in the building blocks of the gene being swapped, a G to a T – the same as with tobacco tar, but in a different hot spot on the gene. Could this be the fingerprint of aflatoxin?

As the paper describing Harris and his colleagues' findings and suggesting such a possibility was about to go to press, a visiting scientist to Harris's lab at the National Cancer Institute mentioned casually that another group, working in South Africa, had also discovered an unusual p53 mutation in liver tumours, but were unsure of its significance or what to do with their findings. Realising that this strengthened their case for a direct link between aflatoxin and p53 in liver cancer, Harris pushed the other group, led by Mehmet Ozturk, to write up their research in double-quick time so that the two papers could be published together, and they came out back to back in *Nature* in April 1991.

The coincidence of aflatoxin exposure and a distinctive p53 mutation in liver-cancer patients soon became apparent in many other warm, humid places with poor storage for crops. But what was going on in the machinery of their cells? Pierre Hainaut joined the quest to find out. Despite his initial unease at being drawn into the tobacco and cancer controversy, Hainaut is a natural-born sleuth, never happier than when he is doing research on the front line, where faulty tumour-suppressor genes are affecting the lives of real people. He has followed the fingerprints of p53 from China to Brazil, Iran to West Africa and South East Asia, and many other countries. With aflatoxin, his research has focused mainly on Mali, The Gambia and Thailand – three countries where liver cancer is a huge problem. Over the years he and others working on this issue, including Harris and Pfeifer, have revealed a devilish relationship between aflatoxin, the ubiquitous Hep B virus and p53, as they co-operate to cause liver cancer.

First the scientists worked out how Hep B virus on its own can lead to liver cancer. A viral gene – known simply as 'x' because for a long time no one had a clue what it did – codes for a protein that has a dual function: one end of the protein encourages proliferation of the liver cells it's infecting; the other end promotes apoptosis, cell death. In this way the virus tries to maintain some kind of balance in the population of infected liver cells. But in so doing, it causes cycles of inflammation, damage and repair to the liver that result in cirrhosis – a liver greatly enlarged and distorted by scar tissue and lumps, or nodules, of regenerated cells.

'These cycles of destruction and regeneration can go on for a while, and sometimes they can kill people – they can die from cirrhosis of the liver without getting cancer,' explained Hainaut. 'But at some point, in the absence of mutant p53, what happens to people with chronic liver disease and cirrhosis is that HBx becomes accidentally integrated into the genome of the liver cells. At that point it loses it pro-apoptotic part, and what remains is just the part that activates proliferation: the cells then escape destruction and are on their way to cancer. This is why cancer develops as a sequel to cirrhosis in the context of wild-type p53.'

Harris's group found also that HBx protein sticks to p53 protein, forming a complex in much the same way as SV40 does with p53. They assumed that in so doing HBx had a similar effect of crippling the tumour-suppressor function of p53, and that this was one of the driving forces towards cancer. However, very recent research by Hainaut and his colleagues in West Africa suggests this assumption is wrong; it has the relationship between the two proteins back to front, for what really seems to be happening is that p53 is blocking the ability of the virus protein, HBx, to trigger apoptosis, while leaving its ability to drive proliferation of cells intact. The crucial point here is that, in real life, the bond between p53 and HBx that transforms the viral protein is only really strong when p53 has the aflatoxin-induced mutation, at codon 249. Then the brakes are off and the liver is especially vulnerable to cancer. 'The risk of having liver cancer for someone who is a chronic carrier of Hep B is about 5–7 times compared to a non-chronic carrier,' Hainaut told me. 'The risk of getting liver cancer with just aflatoxin is very difficult to measure, but is probably no more than twofold. However, the risk of having liver cancer if you have the two is at least 20 times, and some measures suggest it is up to 60 times greater than usual. So it's truly multiplicative.'

The revelation that mutant p53 transforms the function of the virus rather than the other way round has also helped to explain an abiding mystery in African liver-cancer patients. When someone infected with Hep B virus finally succumbs to liver cancer, he or she generally has signs of advanced cirrhosis from years of damage and repair. 'This is the rule in the Western world,' said Hainaut. 'The patient who doesn't develop cirrhosis before liver cancer is really the exception.' But this is not what they have found among patients with liver cancer in Africa, despite chronic infection with Hep B. 'I would say that maybe 15 per cent have cirrhosis beforehand, and then a number of them develop cirrhosis during the proliferation of cancer, as a sort of secondary response of the liver to the inflammatory state, but it does not precede cancer.'

Hainaut's theory is that by blocking the virus's killer function, the mutant p53 prevents the regular cycles of inflammation, damage and repair that cause the scars and nodules of cirrhosis. Thus, paradoxically, aflatoxin exposure can be protective of people with chronic Hep B infection, often for years, until other events in the ordinary course of living render them vulnerable to cancer. 'We could never understand why we have so little liver cirrhosis in these populations. It's something I observed about 15 years ago – we had very few patients with cirrhosis. The common response was, "Ah, you're not looking for them . . . They're not reporting to doctors, so detection is not good . . . The diagnosis is not accurate," and so on. But since then we've done a few cohort studies (which follow a group of people who share common characteristics and lifestyles) and still we find most patients presenting with liver cancer without any trace of cirrhosis beforehand.'

In their studies among liver-cancer patients in Thailand, Hainaut and his team found the same phenomenon – those who had the aflatoxin mutation as well as Hep B infection had no signs of cirrhosis. But though aflatoxin-mutated p53 may be protective of livers in the short term, the case for controlling the offending fungus to reduce the burden of liver cancer is overwhelming, as events in Mali have shown serendipitously. Poring over the cancer register in the capital city, Bamako, very recently, Hainaut and colleagues faced another mystery – liver-cancer cases seemed to be plummeting. In the 15 years since the late 1990s, the rate of new cases had declined by about 75 per cent. They looked for flaws and biases in the records, but could find nothing obvious to explain away the dramatic figures.

On further investigation they discovered that in the mid-1990s, the agriculture ministry had started a programme to prevent aflatoxin contamination of the country's crops. The primary motivation was not public health but economics: Mali wanted to export its crops for animal feed and needed to comply with international regulations. But this has had far-reaching consequences for the man and woman in the street, said Hainaut. 'The first thing is that the contamination of food has decreased; and second, most of the crop production has been diverted towards export, so the diet has changed.' Today, the exposure to aflatoxin of people in Mali is only a tiny fraction of the exposure of people in The Gambia, where rates of liver cancer remain as high as ever.

But there is a downside to this story: as Hainaut and his colleagues predicted, doctors are beginning to see more people with liver cirrhosis in Mali than ever before, as Hep B is still widespread but the factor that keeps the offending viral gene under control – aflatoxin-mutated p53 – is no longer so common.

The great appeal of molecular epidemiology to those involved is that it is often swiftly and directly applicable to real life, and this is the case with liver cancer and mouldy grains. Having worked out the relationship between aflatoxin, p53 and Hep B virus, the scientists find they can read much of what is going on in a person's liver with a blood test. 'The point is that when material is being cleared from the liver it goes either into the bile or the blood. It can't go anywhere else – there's no direct route to the outside world like in the digestive tract or the lungs,' explained Hainaut. 'That means that every bit of DNA from liver cells ends up in the bloodstream. And the liver is such a massive organ that a large part of the free-circulating DNA that you find in the blood comes from the liver.'

The scientists have worked out a method for retrieving that DNA and screening it for aflatoxin-mutated p53. They are also able to monitor the components of the viral genome and to look at what's happening with HBx. Unfortunately, however, no such simple test exists for skin cancer, where the carcinogen, sunlight, leaves an equally clear fingerprint on p53. People just have to be on the lookout themselves for the signs and symptoms of disease.

THE SUN'S FINGERPRINT

I was seven years old when my father was posted to Borneo to run a TB clinic and a general hospital serving the indigenous Dyak tribespeople. The family set sail from Liverpool, and I vividly remember the days on deck, far out in the ocean under a blank blue sky and scorching sun. We all got sunburnt, coming out in big blisters on our shoulders that were too sore to touch and meant we had to sleep, spread-eagled, on our stomachs. Eventually the skin peeled off in long strips like wallpaper, but my younger sister's nose never seemed to heal completely and family photos of our Borneo days show her with a patch of sticking plaster across it most of the time. In those days, the 1950s, and indeed for several more decades to come, we had no idea of the risks we were taking in not protecting ourselves from the sun.

We know now that ultraviolet light (UV) is the main cause of skin cancer, and in the early 1990s Douglas Brash and colleagues at Yale University discovered that it too damages p53 and leaves a characteristic fingerprint mutation. When Brash first started investigating the effect of UV radiation, a known carcinogen, on skin cells in the late 1980s there were three main theories about how it causes cancer. One was that it disrupts the immune system so that it fails to remove damaged cells from the surface of the skin as normal; another that sunlight directly stimulates cell growth; and the third that it damages DNA, knocking out a vital gene or genes.

Several groups in Europe and North America working independently on skin cancer had already discovered that UVB rays are absorbed most readily by the squamous cells, flat disc-shaped cells just below the surface, and slightly less readily by the basal cells deeper in the skin that nevertheless account for the majority of skin-cancer cases (at that time, the effect of the sun's rays on the melanocytes, the cells involved in melanoma – the least common but most deadly form of skin cancer – was still unclear). The researchers had discovered also that UVB radiation damages DNA – and that it does so in a very specific way: it hits always at the point where the two bases, cytosine (C) and thymine (T), are adjacent to each other on a strand of DNA, swivelling them round so that a C is replaced by T, and sometimes two Cs by two Ts. This causes slight but crucial changes in the recipe of the protein the strand produces. This exact mutation isn't seen in tumours anywhere else in the body, where sunlight cannot reach, and is thus considered to be a fingerprint of UVB.

In us, as in all living things, DNA damage and mutation happen all the time as we pass through the mill of life. Mutation is, of course, what drives evolution and adaptation to the environment, so it can be a force for good as well as bad. Brash knew therefore that the fact that UVB causes mutation did not automatically point to this as being the culprit in skin cancer. However, he was persuaded by the general pattern of the disease – and particularly by some evidence from Australia, which together with New Zealand has by far the highest rates of skin cancer in the world – that his best bet was to explore the idea of damaged genes.

Typically, skin cancer develops in middle age and beyond. The Australian researchers had noticed that rates among pale-skinned immigrants – generally from the UK and other parts of northern Europe – who had arrived in the country as adults were lower than among those who had arrived as children. Among the more recent immigrants, skin-cancer rates tended to reflect the rates in their home countries, while those who had been in Australia since childhood had rates similar to other white Australians. This suggested that the insult to a person's skin from UV radiation that had set him or her off on the path to cancer had occurred years earlier, with those who had been in the country longest obviously at greater risk from the powerful Australian sun than those who had spent their childhoods in the clouds and rain of northern Europe.

Brash reasoned that if UV radiation affected the immune system or triggered runaway growth of cells directly, the effect would be much more immediate and transient, and the age at immigration would make no difference to the risk. The fact that there was a difference pointed to a mutated gene, which has a lasting effect, as the most likely cause. His task therefore was to find out which gene was affected. Once again it was like searching for a needle in a haystack: the human genome had not yet been sequenced and everyone still believed it contained at least 100,000 genes, not fewer than 30,000, as we now know to be the case.

Brash and his group followed several fruitless lines of enquiry, looking at known oncogenes, before their luck changed. One day, Arnie Levine appeared at Yale to give a talk in which he mentioned that p53 was found to be mutated in many cancers. Brash heard the talk and suddenly the pieces fell into place: a tumour-suppressor gene was a much more likely candidate for skin cancer than an oncogene because, to cause malignancy, a tumour-suppressor gene requires both alleles to be damaged – or both sets of brakes to fail, to return to our car analogy of Chapter 7. These events could be years apart, thus accounting for the typically slow development of skin cancer often many years after the victim first suffered a bad dose of sunburn.

And there was another intriguing clue that p53 might be involved. People suffering from an extremely rare skin disease called Lewandowsky-Lutz dysplasia develop warty growths, particularly on their hands and feet, that can be profuse and that readily turn malignant when exposed to the sun. Lewandowsky-Lutz disease is caused by infection with certain strains of the human papilloma virus, HPV, which is known to target and destroy p53 protein and lead to cancer in other organs, notably the cervix.

To start the investigation, Brash's group pulled from the medical archive blocks of tissue taken from non-melanoma skin tumours of patients in New York City and in Uppsala, Sweden, where Jan Pontén of the University Hospital had become interested and joined the research effort. All the tumour samples came from sites on the patients' bodies exposed to the sun, such as face and hands. The researchers extracted DNA from each and homed in on the p53 gene, looking for mutations. They found them in 90 per cent of the samples – the great majority bearing the fingerprint of UV radiation and thus capable of producing an active protein with the characteristic modifications to the recipe. Brash and his fellow researchers wrote up their findings in *PNAS* in 1991. But this was before anyone knew how normal p53 worked, and it took another few years – and research by many other groups as well as theirs – for a clear picture to emerge of what happens in the normal course of events when we sit out in the sun, and what can go wrong to cause skin cancer. In essence the picture looks like this: at some point in their lives, most probably during childhood, people who develop skin cancer will have suffered an episode of sunburn which caused a cell or cells to sustain mutation to the p53. We now know that UV radiation causes extensive damage to DNA, but our bodies have an efficient mechanism for repairing it if it's not too serious: enzymes in our cells snip out the damaged stretch of DNA and replace it with a healthy copy. However, the mechanism can fail, and the mutation hot spots are the sites in the genes where, for some reason, the repair process is least efficient.

Peter Hall and David Lane's maverick experiment with the sun lamp on Hall's arm showed that p53 is activated in our skin when we sit out in the sun. Other researchers have since found that in the normal course of events, this activated p53 protein will trigger apoptosis in cells that cannot repair their sun-damaged DNA. But a cell in which p53 itself is damaged will resist apoptosis and will sit around, reproducing the fateful mutation from one generation to the next, until further insults to the skin, often decades later, turn it cancerous.

At this point, says Brash, sunlight delivers a double whammy. A rogue cell with mutant p53 is surrounded by normal cells that, when damaged by sunlight, will respond, as they should, by committing suicide. This gives more elbow room for the rogue cell to spread. 'By inducing healthy cells to kill themselves off, sunlight favours the proliferation of p53-mutated cells,' he explained in an article for *Scientific American* with fellow skin-cancer specialist David Leffell. 'In effect, sunlight acts twice to cause cancer: once to mutate the p53 gene and then afterwards to set up conditions for the unrestrained growth of the altered cell line.' This is the crux of the matter, for it is the expansion of clones of that single mutated cell that precipitates cancer.

'What most people don't realise,' says Brash, 'is that clonal expansion is numerically more important to cancer than making the initial mutation. Exposing yourself to the sun five times will make five times as many mutations. But favouring cell division of the p53 mutant five times will make many more mutant cells. Like compound interest at the bank, this exponential increase soon leads to very large numbers.' The pre-malignant lesion thus formed provides an increasingly easy target for the mutagen – in this case UVB radiation – to inflict the crucial 'second hit' on a p53-mutant cell that will turn it cancerous.

***

What we've been discussing here are somatic mutations – ones that occur by chance in individual cells of the body at some point in a person's life. Sometimes the cell that receives the hit is a sperm or an egg, and the mutant gene can then be passed on to future generations. This is called a germline mutation and it can be very bad news for those who inherit it, because every cell in their bodies will carry the mutant gene.

CHAPTER SIXTEEN

Cancer in the Family

In which we: a) hear about certain families whose exceptional vulnerability to cancer is caused by mutant p53 in all their cells, passed down the generations in sperm or eggs; b) meet Doctors Fraumeni and Li, who first recognised the syndrome that now carries their name.

***

When you're treating a person with cancer, the treatment is for that patient, the prognosis is for that patient, recovery is for that patient, and the family just get support. Here the family is part of the condition – so it's more complex. You are never dealing with just one person and one type of feeling and one type of reaction and one type of personality. You're dealing with the dynamics of a family.

Patricia Ashton Prolla

When he was a small boy growing up on the outskirts of Boston, Massachusetts, John Berkeley took a tumble one day while playing in the garden. The fall raised a bump on the back of his head and when, some days later, it had failed to go down, his parents took him along to the doctor, who diagnosed rhabdomyosarcoma, a rare form of cancer growing in the muscle attached to the bone at the base of his skull. John was four years old, and over the next two years and more he was in and out of Boston's Dana-Farber Cancer Institute (then known as the Sidney Farber Cancer Center). He was treated in the specialist Jimmy Fund Clinic, named after the boy with lymphoma of the guts who had been the poster child for the campaign to establish a research centre into children's cancers back in the late 1940s.

John had surgery to remove a tumour the size of a golf ball from his skull, followed by nearly a year of radiotherapy.

'I recall lying on a table and they would literally tape your body and your head down to the table with what seemed like masking tape, so you wouldn't move,' he told me when I spoke to him over the phone. 'I'm sure it was very difficult to keep a child still in order to perform the radiation treatment, but I just remember it was very unpleasant to be taped down to a table at that age.'

His treatment regime included chemotherapy, which lasted two years and meant daily trips to the outpatient clinic where he was hooked up to an intravenous drip delivering a debilitating cocktail of drugs for eight hours at a time. 'Towards the end of my treatments was the time I started going to kindergarten and I had lost my hair, I was a bald child,' he said. 'Kids can be cruel, as they say, and that was certainly the case for me – I vividly recall going to school and being picked on by the other kids.'

The effort to keep John alive despite a poor prognosis from his cancer team was an emotional rollercoaster for his mother and father too. 'I learnt later on that it nearly tore their marriage apart,' he remembered. 'Back in the 1970s, cancer services weren't even close to being established like they are today. For the families to which this sort of thing happened, it was very . . . I would use the word "barbaric". You were brought in; you got your treatments and you'd go home. There was no friendly atmosphere, no family advocacy support, no sugar coating on anything.'

But the return to robust good health of their plucky small son was not the end to the Berkeleys' troubles. John had a brother, born around the time that his treatment began. In an uncanny rerun of John's story, the boys, then aged 10 and six, were playing with friends in the garden when his brother was hit on the leg by a baseball. The lump from the knock didn't go away, and when his anxious parents took him to the hospital to have it checked out, they were told he had osteosarcoma, a tumour on the bone, and his leg would have to be amputated.

'The next thing that raised the red flag for our family was that my father developed some soft-tissue sarcomas in his early forties,' said Berkeley. The family's medical history raised flags too for the specialists at the Dana-Farber Institute. This was the early 1990s, and the genetic basis of cancer was by now firmly established. What's more, a young Chinese-American oncologist and epidemiologist, Frederick Li, who was part of the team taking care of the Berkeleys, had a particular interest in cancer that ran in families. His name had recently become associated with a cancer-predisposition syndrome that he and Joseph Fraumeni, an epidemiologist working at the NIH in Bethesda, had identified.

In the great majority of cases, people with Li-Fraumeni syndrome, or LFS, are born with a mutant copy of p53 in all their cells. This is a so-called 'germline mutation', which means that it occurs in a sperm or egg cell, and the faulty gene is then passed on from generation to generation by the affected parent, leaving their offspring especially vulnerable to cancer at almost any age.

The Berkeleys were offered genetic counselling. The three who had suffered from cancer were tested and all proved positive for mutant p53. John's brother died in a car accident in 2004 while in the throes of an epileptic fit. His father, after suffering several bouts of sarcoma, died of pancreatic cancer in 2007. In 2010 John was one of a small group of people to set up the Li-Fraumeni Syndrome Association – a web-based organisation offering mutual support to people coping with the intensely lonesome experience of living with the constant threat of cancer. He himself has since had another brush with the disease, being diagnosed with myofibroblastic sarcoma in 2012.

THE DISEASE DETECTIVES

Jo Fraumeni started his career as an epidemiologist with the National Cancer Institute working in a dingy little office above a dress shop in downtown Bethesda, Maryland. Epidemiology was a fledgling discipline at NIH in the mid-1960s and Fraumeni shared the space with just one other person, Bob Miller, also a recent recruit. Fraumeni had studied medicine at Duke University, North Carolina. During his residency at Memorial Sloan Kettering in New York he had discovered that he had an eye for patterns of disease and a special interest in looking beyond the patient seated before him in the clinic to the context in which he or she had become sick – the environment in which his patients lived and worked and the families they came from – for clues to what ailed them. Reaching a decision point in his career as a physician, he might well have joined the armed forces and been sent to Vietnam, but he chose instead to specialise in epidemiology at NIH.

Bob Miller was a paediatrician, but he had a degree in epidemiology as well. The two men struck up a firm friendship. 'Bob was kind of an iconoclast,' commented Fraumeni. 'I remember one of the first things he did was to show me a picture on his wall of a cover from the *New Yorker.* It was of seagulls standing in a row along the roof of a house. All except one bird, standing a little apart from the others, were facing in the same direction. The ones looking forward, he told me, are considering the diagnosis and how to treat the patient. "We should be like this one," he said, pointing at the odd man out, "looking sideways to see what else is going on in the patient, in the community, in the environment, in the family."'

Fraumeni still has that picture and he pulled it from a sheaf of papers he had gathered to show me when I visited him at the NIH's National Cancer Institute in the summer of 2012. He was just about to retire, at the age of 79, as Director of the Department of Cancer Epidemiology and Genetics, which he had founded 50 years earlier and which had grown under his leadership from its humble roots above the dress shop into a huge and prestigious unit housed today in purpose-built laboratories and offices on an elegant modern campus outside Bethesda. After a sweltering hot day the sky had gathered thunderclouds which broke just as I arrived. As we sat talking in his book-lined office with its myriad framed certificates and awards dotting the walls, hailstones slashed at the window and leaves stripped from the trees danced in the wind behind the glass. After a pause for reflection, Fraumeni, a reserved man with a spare frame and large owlish spectacles, who chooses his words carefully, said, 'I was very fortunate to have met Bob. He was what I call a kindred spirit.'

Miller sparked his curiosity in childhood cancers, about which almost nothing was known beyond the fact that children with Down's syndrome are especially prone to leukaemia. With this one example of a link in mind, the two epidemiologists decided to look for other patterns in the occurrence of paediatric cancers for clues to what might be triggering the disease at unusually young ages. Cancer of any kind is relatively rare in children, so in order to get enough cases for patterns to become apparent, they needed to scour the records of multiple hospitals. Their first investigation was into Wilms' tumour, a cancer of the kidneys for which their searches turned up 440 cases. These were associated with a 'constellation of anomalies' that included genitourinary defects and mental and physical retardation, as well as aniridia, or absence of the iris, in the eyes of some affected children. This condition is so rare that the six cases of aniridia they found among their sample were 'off the wall statistically' – showing a frequency more than 1,000 times what would be expected in the general population – and therefore strongly suggestive of a link with the tumour.

'We were excited by all this and we went after all different kinds of childhood tumours,' Fraumeni told me. 'We found not only associations between the tumours and the anomalies, but between tumours and tumours. Kids with one cancer sometimes developed a second cancer, unrelated to the first. We found that some of them were related to treatment – radiation and chemotherapy – but a lot of them were not, so it led us to conceptualise the notion that genetic susceptibility can result either in multiple tumours in the same individual, or multiple cancers scattered over the family tree.'

By this time, Fred Li, also a paediatric oncologist by training, had joined their tiny unit. 'Fred was a wonderful guy,' commented Fraumeni. 'He was a prince. Easy to work with, very bright, very modest – he added a tremendous amount to our group.' (Fraumeni has a quaint habit of describing any man he admires – including Alfred Knudson, whose work on retinoblastoma, you will recall, led to the discovery of tumour-suppressor genes – as 'a prince'.) The story of the syndrome to which he and Li gave their names began with a survey of children with adrenocortical tumours – a type of cancer that involves the outer layer of tissue of the adrenal glands. These tiny organs sit atop the kidneys and produce the hormones that control many of the vital functions of living, such as heart rate, blood pressure, the fight-or-flight response to stress, growth and sexual characteristics. 'Adrenocortical tumours are very, very rare,' said Fraumeni, 'so we had to go to about 10 hospitals to get 21 cases. And two of them developed brain tumours . . . Unusual!' he said, pausing for emphasis. 'Then when we finished that survey, I got a call to say that a third case had developed a brain tumour and the family was riddled with sarcomas. That suggested to me there was enemy action . . . Three cases like that was unusual to say the least. So there were all these little clues coming up that something was going on.'

The eureka moment came when a family was referred to Fred Li in which two young cousins had rhabdomyosarcoma (the muscle tumour that afflicted John Berkeley, whose story opened this chapter); the mother of one child, still in her twenties, had breast cancer; and the father of the other had acute leukaemia. 'It was explosive, you know, so much cancer at the same time,' commented Fraumeni. The two doctors homed in on the family, taking a detailed medical history across the generations that revealed an abundance of cancers, sometimes with multiple tumours in the same individual. 'That was it – we knew there was a syndrome.'

Thoroughly caught up in their detective work by now, the two returned to their multi-centre survey of childhood cancers and pulled out the cases of rhabdomyosarcoma for further investigation. Among these they found another three families in which the children's tumours were associated with cancer in their still-young parents. Convinced they had uncovered something important in a field that still had little clue as to how and why children should be afflicted by a disease normally considered to be caused by the ravages of age and long-term exposure to harmful environments, they rushed to prepare a paper for the *Annals of Internal Medicine* as quickly as they could. 'The thing that took the longest time was deciding what to call our syndrome,' laughed Fraumeni. 'Fred was more conservative than I was and we ended up with a question mark after the words "a familial cancer".'

The year was 1969, way before the genetic basis of cancer had been convincingly demonstrated, and most of the seagulls on the roof were looking to viruses. Li and Fraumeni's paper with its suggestion of a familial syndrome – by definition genetically based – was received with frank scepticism. A handful of familial, and therefore hereditary, cancers were already known, but they tended to give rise to the same tumour type in related cases – either breast or colon or ovary, for instance – and to affect adults. 'Our syndrome was unusual in that those affected were all children and very young adults – and a *bizarre* array of tumours . . . Every conceivable cell type, from leukaemias to gliomas to sarcomas to adrenal tumours to breast cancer. And there were some other strange tumours like choroid plexus tumours, which are in the lining of the brain where the cerebrospinal fluid circulates. These tumours have come up so often in LFS that they're almost pathognomonic (i.e. a defining characteristic of the disease). It's like adrenal cortical cancers in children – when you see them, you think LFS.'

The sheer variety of tumour types was what puzzled the sceptics. Some suggested that a virus, transmitted from mother to child during pregnancy and birth much like HIV, was behind the family pattern. Others said there was no real pattern and that what they were seeing was 'just the play of chance'. But while, as good scientists, they had to keep open minds, Li and Fraumeni were pretty convinced that a gene or genes were at the root of the clusters they had observed. They decided to keep a close eye on the four original families in their survey, and over a 12-year period they saw a further 16 cases of cancer develop – all of them part of the same constellation of tumours that characterised their syndrome.

During this time two other groups had joined the field of familial cancer epidemiology: one led by Louise Strong at MD Anderson in Houston and the other by Jill Birch at Manchester University in England. Together they provided much new evidence in support of Li and Fraumeni's genetic hypothesis – and, incidentally, were responsible for the name Li-Fraumeni syndrome, or LFS, creeping into the medical literature and common parlance in the 1980s. Almost another decade was to pass, however, before mutant p53 was identified as the gene bringing grief to the unfortunate families.

HOMING IN ON THE RESPONSIBLE GENE

In the late 1980s Fred Li, then Head of Cancer Epidemiology and a practicing oncologist at the Dana-Farber Institute in Boston, was joined on the staff by a young Canadian named David Malkin, trained in paediatrics at Toronto's Sick Kids Hospital, who was looking to gain some experience in children's cancers to take back to the clinic in Canada. But the time he spent with Li – whom he describes as 'a gentleman and a scholar; not physically imposing but with a real presence, and extraordinarily knowledgeable' – gave Malkin an appetite for research, and he decided to pursue a career as a scientist alongside his work in the clinic seeing patients. He got a position as a postdoc with Steve Friend, the scientist who had discovered the first tumour-suppressor gene, Rb, just three years earlier. Friend had recently left Weinberg's lab to set up on his own at Harvard, and over a casual meal in a restaurant he and Malkin brainstormed ideas for a research project to work on together. 'Steve explained that he liked working with Fred (Li) and they were interested in exploring the genetics of this weird syndrome. I knew relatively little about LFS at the time,' commented Malkin, 'but it sounded intriguing.'

With a vast arena to work in and no clues to guide them, the two decided to start their search for the faulty gene by looking at Rb – if only because, with an inherited condition, a malfunctioning tumour suppressor that would leave carriers unprotected against cancer seemed like the most obvious candidate. Rb was familiar territory to Friend and there was little to excite their interest in p53, which had yet to emerge from the doldrums of being miscast as an oncogene. Coincidentally, the two scientists drew a blank with Rb just as big things began to happen again with p53. In 1989 Suzy Baker and Bert Vogelstein in Baltimore revealed that wild-type p53 was in fact a tumour suppressor, and later that same year scientists working with mouse models in Toronto published a paper describing the multiple tumour types that developed in animals with mutant p53. The constellation of mouse tumours didn't exactly match what oncologists were seeing in families with LFS, said Malkin, but it was a dramatic enough demonstration of p53's broad effect to make the researchers switch their focus to this newly revealed tumour suppressor.

Working with material provided by members of LFS families seen by Fred Li and Louise Strong in their cancer clinics, the two scientists set about isolating p53 to look for possible mutations. In those days it was an almost infinitely laborious process of cloning and sequencing stretches of DNA one after another until they found their target gene. 'A lot of sequences didn't work, and it was well over a year before we had our first hit that we knew was real,' commented Malkin. 'I remember I looked at the gel; I sequenced the gene; I compared it against the normal p53 and there was the mutation. We were very lucky because the first one that popped up was at codon 248, which happens to be about the most common mutation we have in p53. That was fortuitous, because if it had been in some slightly more obscure site we may not have paid so much attention to it.'

He and Friend knew there were months of hard work ahead if they were to confirm the finding and look for more mutations but, once they realised they were on the right track, the whole lab went into overdrive. 'That's where you turn your alarm clock off and you're just in the lab the whole time because you're having a lot of fun!' said Malkin with a grin. 'That summer of 1990 was very exciting. We were in a brand new building, it had a beautiful view over Boston Harbour . . . There was no good reason not to enjoy yourself and the whole lab was really on fire.'

Having finally put the picture together of what was happening in people with LFS, Malkin and Friend hammered out their paper for *Science* at a nearby pizza restaurant, the table scattered with handwritten pages amid the plates of half-eaten food, and the two preoccupied scientists oblivious to the comings and goings of other diners. This they knew would cause a stir because, although LFS appeared to be extremely rare, identifying mutant p53 as the root cause of a problem with such diverse manifestations confirmed the gene's central importance in human cancer. Not just in lab dishes and genetically engineered mice, but in real people like us. Here too was a means of identifying people at risk in cancer-prone families who, with intensified screening, could be treated for lesions before tumours had a chance to develop.

Over the years mutations have been reported in many different sites in the p53 of people with LFS, just as they have in so-called somatic cases (people who have developed a mutation during life rather than inherited a mutant gene at birth). The most common ones are those that affect p53's ability to attach itself to DNA in order to switch on other genes (i.e. to act as a transcription factor). Researchers have discovered too that girls born with LFS have a lifetime risk of developing cancer of more than 90 per cent, while the comparable risk for boys is around 75 per cent, with breast cancer most probably accounting for the higher risk to females. Though most oncologists recognise a typical pattern to LFS, said Fraumeni, 'One of the things that I've learnt in studying these families is that the susceptibility is not just limited to the so-called classical tumours – brain, adrenal, breast, sarcomas. It really is almost across the board; almost every tissue, I think, has an increased risk.'

LIVING WITH LFS – LUANA LOCKE CONTINUES HER STORY

These were the odds that Luana Locke faced when she learnt that her family had Li-Fraumeni syndrome, giving her an explanation at last for why lightning kept striking in the same spot. When Luana's mother, sister and aunt died of cancer in the early 1970s, LFS was barely recognised outside of academic circles and had certainly not been named. It was not until her Aunt Rina's youngest daughter, Jessica Zendri, in Italy developed bilateral breast cancer and an adrenocortical tumour that a geneticist in Milan discovered the germline mutation in p53. He suggested that Luana be tested for the same genetic defect back home in Canada, and he recommended to her someone with special expertise,

David Malkin, now working again at Sick Kids Hospital in Toronto.

After all she had been through, Luana was not surprised to receive the news that her test result was positive, she said. But then all of a sudden the focus was shifted away from her. 'All I remember hearing was, "You have a 50/50 chance of passing this on to your child." Now it was all about my son: does he have a mutation, does he not? So it was just about getting him tested, because I needed that knowledge; I needed to know.'

Subsequently, as she and her husband Paul sat through genetic counselling, and consulted again with Dr Malkin, the implications of her diagnosis and the possibility that Lucas, now four years old, also carried the mutant gene, were carefully explained. 'Everybody we spoke with did their due diligence,' she commented. '"These are the things you really need to consider: the potential for false positives and false negatives; that this is a decision you're making on behalf of your child, so you'll need to think about at what stage you tell him. What if he's the type of person who, in adulthood, says he wouldn't have chosen to be tested; he wouldn't have wanted to know?" All of these things they were telling me . . . but I had already stopped listening. I sat there and I went, "Aha, yeah, okay, I have to think about that." And all along an inner voice was just saying, "He's going to be tested; I need to know."'

What did Paul feel? 'I'll admit, there wasn't a choice there, I would have played the mother card, because I'm that type of person. There's no way somebody's going to hold a secret that's just within my grasp and me not know about it. That would have killed me. And secondly, I'm the eternal optimist. For me it was, "Huh! He's not going to have the mutation! Just give me the results so we can close that chapter and move on."'

After a period meant for reflection, the Lockes took Lucas along for testing. But none of the messages had sunk in and Luana was not prepared for the result. Sitting across a small table from me in the corner of a busy restaurant in Toronto 12 years later, she vividly recalled the meeting at the clinic. 'The genetics counsellor who was working with Dr Malkin's programme came with this manila envelope. She sat us down, Lucas was there too, and she started talking *science,* right? She was talking about the gene, and this strand and that strand, and I was just following along pleasantly. I was like, "Okay . . . she's just stringing us along and this is going to end with: '. . . but there's nothing there; clean bill of health!'" And so when she said, ". . . and we did find . . ."' Locke's jaw drops as she recreates her moment of shock. 'I thought, "I'm sorry, *what?*" I didn't say it out loud, but I remember thinking, "Okay, wait a minute, I was not prepared for this. That's not the way this story was going to end for me."

'I think the poor woman who was telling us this stuff must have been thinking, "When is this woman going to start getting that I'm going somewhere with this story? Because she's still smiling along. . . ."' Only then did all the questions Luana and Paul had been advised to consider in advance flood their minds. 'I stopped and I thought of every single one of them. Oh my gosh, what if he doesn't want to know? How am I going to be the keeper of this information and knowledge? And am I going to treat him differently? Of course I am . . . We just felt extremely protective.' That night she and Paul took Lucas into bed with them – a practice they had never indulged before – and held on to him tightly.

With time, the Locke family came to terms with LFS. They fell into a routine of regular three-monthly visits to Sick Kids for Lucas to be screened, and in due course allowed themselves to consider having another child. Traumatised by the regular tragedies in his extended family, Paul was only ready to go ahead provided they agreed to screen the unborn baby and consider termination if he or she carried the mutant gene. But abortion was never really an option, commented Luana. When she got pregnant and began to think of screening, she realised she could never go through with it. She was haunted particularly by the question: had such a test been available when her mother was starting a family 47 years ago, would she have chosen to abort Luana and her siblings?

So Lucas's little sister Juliet was born in 2006. Once again Luana believed she would be okay. 'After all,' she said, 'two kids, 50/50 chance . . . I'd had the 50 per cent already.' But once again her optimism was misplaced. Paul and she were given the news that Juliet too is a carrier of the mutant gene, and they will need to help their daughter, as they are helping their son, face up to life in the shadow of cancer.

As John Berkeley has found, coping with LFS can be an isolating experience. Until recently, however, the condition was considered to be mercifully rare. As of 2011, only about 500 families worldwide had been reported in the literature with germline mutation of p53. But that was before Dr Maria Isabel Achatz revealed what was going on in southern Brazil.

CHAPTER SEVENTEEN

The *Tropeiro* Connection?

In which we learn about Li-Fraumeni families with an atypical p53 mutation thought to have been introduced to Brazil by an individual settler from Europe in the mid-19th century.

***

Nothing in life is to be feared, it is only to be understood.

Marie Curie

São Paulo is a giant sprawl of a city with extremes of wealth and poverty. Rough wooden shacks cluster in hidden spaces at the foot of skyscrapers of glass and steel, and here and there destitute people huddle in doorways on streets along which their fellow citizens in designer clothes hurry to work or shop or meet up with friends. Everywhere, huge tropical trees, as if in dogged defiance of the relentless spread of concrete and tarmac, buckle the pavements with their roots, drop their blossoms on passers-by and harbour chattering birds in their leafy crowns. This is the city where Maria Isabel Achatz lives and where she began to uncover an extraordinary story of familial cancer as she worked at the huge, modern A C Camargo Hospital. And it is where I visited her in the late summer of 2012 to hear her story.

A C Camargo, the largest specialist cancer centre in Latin America, was purpose built in 1953 on the site of a former Japanese temple a short distance from the busy boulevards of the city centre. This is the district of Liberdade, a network of narrow, shabby streets lined with small shops, businesses and open-fronted diners arranged on hills that afford sudden views across São Paulo, out on to green squares and old churches and beetling traffic. Liberdade is home to the biggest community of Japanese people outside Japan. With its complex of modern research laboratories and training facilities alongside the imposing white building of the hospital itself, A C Camargo dominates the district. Every year some 15,000 new patients seek diagnosis and treatment here. Around one in 20 of them comes from outside Brazil, attracted by the fact that A C Camargo's oncologists are specialists in many different aspects of cancer and are highly skilled.

Achatz is an oncogeneticist, a relatively young specialism in medicine that is concerned with cancer cases in which the faulty genes are inherited rather than occurring by chance during the course of life. She grew up in a cosmopolitan family in Rio de Janeiro, the daughter of an economist father and housewife mother, and the youngest of six children. On finishing high school she went to university in Paris to study art and design. But her plans for a career in the arts came unstuck when she joined a bunch of her university friends on a visit to India. The group spent time on the Kashmir border, camping in a desert area close to the site of a leprosy colony where the people were living as outcasts in caves in a stony hillside. Their plight, and their extraordinary resourcefulness, made a deep impression on Achatz. But it was a chance meeting with Mother Teresa, whose religious order ran the colony, that made her decide on a career in medicine instead. 'It was an amazing encounter and I thought, well, I just have to go back and do something more worthwhile.'

Achatz returned to Brazil and eventually to medical school in São Paulo – a training that was interrupted briefly by the birth of two of her children. Family life has always been important to her and she says, 'I decided to go into genetics because I wanted to work with families. Mostly I wanted to work with cancer genetics, which was an emerging field.' A C Camargo has one of the busiest oncogenetics programmes in the world, and during her first year there, while she was still a trainee, she saw 30 families with what she believed was LFS.

'It really struck me because this was considered to be a terribly rare syndrome. There were only 280 families described in the world literature at that time – and I had 30. So I thought, either I'm over-diagnosing or something unique is happening here,' she told me as we sat talking in her small, windowless office behind the lab in one of the research buildings. Maria Isabel is tall, slim and effortlessly elegant, with long brown hair which she occasionally ties back into a swinging ponytail. Dressed casually in white shirt and slacks, she sat behind her desk, always prepared for a phone call from one of her patients or a colleague or student.

There are more than 70 cancer-predisposition syndromes, she told me, but she thinks she has an eye for spotting LFS, having seen her first family with the syndrome while she was still a medical student at another hospital in São Paulo. 'This patient came just for clinical follow-up, but after the consultation she said, "Well, actually I am a cancer survivor. I've already had five different cancers." I said, "You mean metastases?" And she said, "No, no, five different cancers." Then she listed all the different tumours she'd had and she said, "This is something really common in my family – we have so many cancers . . . It just happens, and we get better afterwards."

'I was so fascinated that I studied and studied, and when I came to this hospital as a trainee, again, the first patient I saw was a Li-Fraumeni patient. They just kept coming and coming, till I had by the end of the year these 30 families with LFS.'

One of the ways scientists communicate their findings to each other is through the medium of posters – short summaries of their research projects that they are invited to pin up on special display boards in side rooms of scientific meetings. In spare moments, delegates can wander around reading the posters at their leisure. Achatz's boss at A C Camargo encouraged her to submit a poster on her LFS families to a cancer conference scheduled for France in the winter of 2002, and when it was accepted she and he flew together to Paris to attend. Achatz recounted how she found herself standing before her poster discussing her work with a tall, earnest, bespectacled scientist whose name she didn't know, when an agitated official came to tell him he was due on stage any minute as chairman of the next session. As he hurried away, Pierre Hainaut handed her his card, 'and I realised I had been talking to one of the top scientists of WHO's centre at IARC,' she laughed.

As keeper of the p53 mutation database, Hainaut was well aware of LFS and its supposed rarity, and he was fascinated by Achatz's story. He invited her to visit him at his lab in Lyon before flying home from the conference, and after much negotiation the two agreed to collaborate on research into the Brazilian phenomenon. Maria Isabel had never seen herself as a research scientist – 'I was very happy being a medical doctor, taking care of my patients and trying to detect their tumours early,' she told me. Besides, at the time of the Paris conference, she was pregnant with her fourth child and fully intended to drop back into the carefully constructed routine of family and hospital life she had left. Encouraged by Hainaut, however, she did a PhD and today she combines research, and the supervision of a bunch of students, with a busy medical practice at A C Camargo.

Oncogenetics is a particularly demanding specialism, she told me, 'because I bring my patients the bad news that their families will have to take special care of themselves forever. I'm not someone who performs surgery and heals people. I'm only able to offer a hand and walk together with them; I'll never be able to withdraw their suffering.'

She tells the story of three siblings, a girl and two boys, all young adults, who decided to join her research programme having watched their mother cope for years with multiple episodes of cancer. After counselling the three and making sure they understood the implications, Achatz took blood samples and sent them off for analysis. She was startled to learn from the mother the following week that one of her sons, a young man with a PhD in engineering and a good job, had left the clinic after the consultation and, without waiting for his results, quit his job, left his wife and two children, and had a vasectomy. He wanted to live life to the full while he had the chance, he said.

'The thing is that the results came back soon afterwards and he wasn't even a carrier,' said Achatz, shaking her head at the needless tragedy of the story. 'A very educated person . . . We took all the time we needed to counsel him. He understood it well. But this shows how big is the impact of having a genetically inherited disease: your education, your IQ, your background have nothing to do with how you react. Getting the information is just such a hit that no one can tell how you'll take it.'

At their first meeting in Lyon back in 2002, Hainaut asked Achatz to collect blood samples from her cancer patients in São Paulo and bring them to IARC for analysis. After extracting the DNA, the two scientists started by isolating and sequencing the sections of p53 in which the great majority of significant mutations – including the ones most commonly associated with LFS – occur. But out of 45 supposed LFS patients from whom Achatz had drawn blood, they found only three with mutations. Achatz was mortified. 'I said, "Well, I was wrong. I over-diagnosed. I don't have anything unusual here in Brazil."' But Hainaut wasn't so easily put off. They must now sequence the whole gene, he said, including the sections outside the usual hot spots, at either end of the gene. Sure enough, they found that a good proportion of the individuals carried mutant p53, with a mutation at codon 337 being most common.

Codon 337 falls in a part of the gene responsible for how the protein folds, and therefore how it interacts with other proteins in the cell. Hainaut was aware that this same, unusual mutation had been described just the previous year, 2001, in association with adrenocortical cancer (ACC) – significantly also by scientists working in Brazil. ACC, you will recall, is the rare childhood cancer that Li and Fraumeni had investigated when searching for evidence of a cancer syndrome that runs in families. But the Brazilian scientists, led by Raul Ribeiro, claimed there was no evidence that the 337 mutation predisposed to anything other than ACC in children. This was not part of a wider cancer syndrome, they said, and in their paper for *PNAS* they gave a convincing explanation of why the risk should be limited to the one organ: that 337 was a 'conditional mutant' that caused the protein to misfold only at a certain pH level (a measure of acidity in the cells), a level found in the adrenal glands. They suggested, moreover, that an agricultural pesticide widely used in southern Brazil might be responsible for causing the mutation in the sperm or egg cells that resulted in children being born with faulty p53.

'This was a story that kept coming up,' said Hainaut, 'and in my first contact with Maria Isabel we did discuss the possibility of the pesticide.' Anxious to investigate further the connection between the 337 mutation and a wider spectrum of tumours than Ribeiro's group had found, he flew to São Paulo to do further work with Achatz in 2005. She had recently begun collaborating with Patricia Prolla, who runs a medical genetics clinic at the big general Hospital de Clinicas de Porto Alegre, in the south of the country, and who was also seeing patients with apparent LFS. Together the three scientists travelled to Ibiuna, a small town on the outskirts of São Paulo where one of Achatz's most open and accepting patients had promised to set up a family meeting at her home. The three found 26 members of the family crowded into the elderly matriarch's kitchen, and over cakes and coffee round the kitchen table they listened to their stories the whole afternoon.

'They'd seen so many deaths in their families and nobody had ever said there was something to be done, so they just thought, "Well, that's our family,"' Achatz told me. 'Many of them spoke about being cursed.' But the information the three scientists gave the men and women gathered there – representing three generations of the family – about LFS and the screening procedure that could determine who among them was a carrier of the mutant gene and who was not, kindled a spark of hope that challenged their fatalism. Achatz, Hainaut and Prolla finally left the house with blood samples from almost everyone in attendance.

This was just a small grouping of the large extended family that lives in and around Ibiuna, however. Word of the meeting with the A C Camargo team soon spread among those who had been reluctant to delve too deeply into their painful history, and many more decided they wanted, after all, to hear why they were so cancer-prone and what could be done about it. A few weeks later, around 85 members of the family travelled to A C Camargo for a meeting with Achatz. 'That was in 2005, and I still see them almost every week – one of the families comes here,' she commented. 'One cousin brings another cousin – they just keep coming.'

WHERE DID IT START? HOW FAR HAS IT SPREAD?

As the extent of the problem in the Ibiuna family became apparent, Achatz and Hainaut began to ask the questions: where did the mutation originate? And how far back did it go? With the help of a local priest who had access to parish records of marriages, births, baptisms and death, the researchers were able to trace the family back to the early 19th century and to draw up a family tree with several hundred people across eight generations. Wherever possible they added details about who had suffered from cancer and of what organs. The trail of disease goes back a long way and one of the family members suggested it could be part of their *tropeiro* heritage.

The *tropeiros* were travelling traders, muleteers who supplied the early Portuguese settlers in the mining and farming communities dotted around southern Brazil with all manner of goods, as well as carrying mail and news from the outside world. The traders travelled great distances, were on the road for long periods of time and likely had girlfriends all along the route. In a scenario reminiscent of the AIDS pandemic, someone with a mutant gene could have passed it on in much the same way as HIV was disseminated along the trucking routes of Africa and Asia. Such a genetic fault is known as a 'founder mutation' – one that is introduced to a population by a single immigrant and can be traced back to an individual, a common ancestor, through the DNA. It was an idea that begged to be explored because, like a stone thrown into a pond, the ripples from a founder mutation can spread far and wide. How far had 337 mutant p53 spread in Brazil?

It has turned out to be remarkably widespread. When Prolla and Hainaut did some research in Porto Alegre – also on the trade route of the *tropeiros* and their mules – to investigate the prevalence of the mutation in the general population, they came up with startling results. Prolla's particular interest is breast cancer, and she was involved in a study of prevention strategies that recruited several thousand healthy volunteers from poor suburbs of Porto Alegre to test mammography. She and Hainaut secured permission to test the blood of 750 of the women. It was a shot in the dark; they had no idea what to expect, but they found two samples positive for the 337 mutation. 'The frequency was *significantly* higher than we expected,' said Prolla. 'In studies that have been published in Europe and the US, the frequency of a germline mutation in p53 is about one in 5,000 individuals in the general population. If we really think that the frequency of the 337 mutant is somewhere around one in

300 or 400 in the south and southeastern regions of Brazil, for one single mutation, this is much more common – at least 10 times more common – than any other germline p53 mutation anywhere else in the world.'

The extremely high prevalence was confirmed subsequently by a screening programme of newborn babies in the southern Brazilian state of Paraná. Of 171,649 babies whose parents agreed to the test, 0.27 per cent were found to be carrying p53 with a mutation at codon 337. Such findings were disconcerting: could this mutation really be a cause of the cancers, or was it just a red herring – a harmless variant of the gene – the scientists wondered? The evidence that the mutation is indeed harmful was strong, however. Besides the fact that the risk of ACC in carriers is 10–15 times higher than 'normal', women with the mutation tend to develop breast cancer at least 10 years earlier than those without.

The 'common ancestor' hypothesis was confirmed when the scientists found that every LFS family member who had tested positive for the 337 mutant in Brazil had an identical version of p53 – there was no variation to suggest that the same mutation had occurred spontaneously and independently a number of times. This remains true for more recently identified cases also. The lack of variation tells us that the original case was relatively recent, said Hainaut. 'It's not an old founder effect that goes back a thousand years or so, or you would expect to see some kind of "drift" in the non-coding region of the gene.'

There's a postscript to the story of Prolla and Hainaut's prevalence study in Porto Alegre. Prolla followed up their research by calling in the two women who had proved positive to discuss their family histories. Cancer was all too familiar to them, the women reported. And to everyone's surprise they discovered, while drawing up family trees with Prolla to track the disease across the generations, that they were related: both reported the same distant cousin who had cancer.

Not long after these consultations, the baby niece of one of the women developed ACC. Because of the awareness of symptoms raised by Prolla's work with the family, the baby was brought promptly to her clinic, when her tumour was easily treatable. But her parents were reluctant to consent to surgery because of their religious beliefs – they were Jehovah's Witnesses, whose faith prohibits blood transfusions. Much to their relief, the surgeons removed responsibility from their shoulders by securing a court order for treatment to save the child's life, and in the event no transfusion was needed. Today she is a healthy little girl who visits Prolla's clinic and the hospital regularly, along with her father and aunt, for her routine screening, as all three are carriers of the p53 mutation.

A MORE COMPLICATED STORY

The *tropeiro* connection is intellectually appealing, and the Ibiuna family can trace their roots back to an ancestor, an immigrant from Portugal, who made his money as a travelling trader and bought land in the Ibiuna area to grow grapes for wine in the early 1800s. But in the end it can never be proven that he was responsible, and Hainaut felt from the beginning that the true story was probably more complicated. He has another hypothesis about the origins of the founder mutation that he feels is equally plausible and as a result has spent nearly two years investigating alternative possibilities, often accompanied by Achatz, as time permitted. The mutation is known to be concentrated in southern regions of Brazil, and the trail has taken the two to towns and settlements along the old trading routes between São Paulo city and Porto Alegre, 860km (530 miles) away in the state of Rio Grande do Sul.

Codon 337 is a vulnerable site on the gene for mutations, explained Hainaut, and to investigate possible carcinogens he and Achatz travelled to a grim industrial town where the population is exposed to pollution from heavy metals, sulphur and other chemicals seeping from dirty mine dumps. But they found nothing of particular interest there. They visited a coastal town which was a popular entry point to the country for European immigrants in centuries past, and frequented by sailors from round the world, but again their amateur sleuthing turned up nothing.

The third idea they followed took them to Araranguá, 210km (130 miles) north of Porto Alegre, which had been settled by the Portuguese since the early 1700s. They spent hours poring over the old archives in the town hall and here they felt they might be on to something. The records showed that the roads used by the *tropeiros* had started from Laguna, another small city some 100km (60 miles) to the north. They had been built originally on the orders of King João V of Portugal as part of his strategy to secure the territory inland, between the coast and the central plateau, for his crown. At the time, the area was occupied by indigenous peoples, with occasional incursions by Spanish troops from the Río de la Plata to the south – what are today the countries of Argentina and Uruguay. The king's desire to control the area was driven largely by the fact that rich seams of gold were being mined further north, and he wanted to occupy this untamed land in order to prevent the Spanish from moving in from the south and seizing the mines. The people of Laguna were therefore recruited to build a route to carry soldiers and settlers as fast as possible towards the escarpment.

The archives told a story very similar to that of America's conquest of the West, in which the settlers wiped out the indigenous peoples they encountered and established their own communities in villages all along the route as they carved out roads to the interior. The groups who opened the road consisted of a small officer class of white Portuguese accompanied by a large contingent of slaves. The whites are likely to have been the main political force in the new settlements strung out along the Ararangá road, said Hainaut, and to have maintained their elite social status by marrying among themselves.

'This meant that a very small group of people originally from the same city became, within about 15 years, the seed of all the white population of that area. I believe one of them had the mutation, and through intermarriage it reached a high prevalence in the community from very early on. These very particular historical and demographic circumstances could explain how the mutation got so firmly established despite having a negative effect.'

By the middle of the 18th century the road stretched in a continuous track from Sorocaba, inland from São Paolo, to Porto Alegre, and soon became a busy trading route. Trade consisted largely of cattle being taken to market in Sorocaba from the south, and goods of all sorts being brought down by mule on the return journey. It was a round trip of some six months for the *tropeiros,* who would have had many stops, and likely liaisons, along the way – perfect conditions for spreading a genetic mutation.

Again, the theory cannot be proved and Patricia Prolla, for one, is wary of all possible explanations advanced thus far for the high prevalence of the mutation in Brazil: they leave too many questions unanswered, she believes. But Hainaut remains intrigued. When the road reached the escarpment, it branched north to São Paulo and Rio de Janeiro, and south towards the small settlement that grew eventually into Porto Alegre. The leader of the team that drove south was Francisco de Brito Peixoto, a famous figure who appears in history books as the first explorer of Rio Grande do Sul, the southernmost state in today's Brazil, and the founder with his father of the coastal city of Laguna. He died in 1735 and is said to be buried beneath the altar of a small chapel in Laguna; Hainaut dreams of getting permission one day to disinter his bones and to sample his DNA as the possible first carrier of p53 with the 337 mutation.

AN UNUSUAL PATTERN OF DISEASE

A mutation that kills people prematurely usually dies out within a few generations as it inhibits child-bearing and thus the passing on of the faulty gene. It's one of the ironies of the Brazilian form of LFS that it is more persistent and widespread than the classic syndrome because the 337 mutation is somewhat 'weaker'. Carriers of this mutant have a lifetime risk of developing tumours of around 60 per cent and they tend to get sick later in life. Thus around one in four carriers of the Brazilian mutant remains cancer-free until at least their 30th birthday, whereas the comparable figure for those with classic LFS mutations is only one in two. The spectrum of tumour types is somewhat different also, with leukaemia and sarcomas being less common among Brazilian patients, but ACC and a tumour of the central nervous system called choroid plexus carcinoma being more prevalent in young children.

Two remarkable observations may hold clues to why the pressures of natural selection have failed to eliminate 337, as they would be expected to do. One is that those carriers in Brazil who don't get cancer are likely to go on to a healthy and vigorous old age, perhaps well into their nineties, suggesting that there are some counterbalancing advantages to having the mutant. Another is that carriers of the 337 mutant do not get the cancers associated with viral infection, such as cervical cancer or Hepatitis B-associated liver cancer. Perhaps the mutation protects carriers from viral infection or from some other disease that plagued the early Portuguese settlers, giving them a survival advantage, suggested Hainaut. This is just wild speculation at the moment, but unlike the theories about the mutant's origins, at least it is testable in the lab.

If the prevalence rates of the 337 mutation found in the screening programmes of newborn babies in Paraná and women attending mammography services in Porto Alegre are truly representative, several hundred thousand people in southern Brazil are living with mutant p53 and are at high risk of developing multiple cancers. Precious few are within reach of the specialist genetic counselling and treatment services of São Paulo's A C Camargo and the Hospital de Clinicas de Porto Alegre. For some of those who are, however, the regular check-ups put their minds at rest and allow them to live normal lives. Others are unable to shake off the spectre of cancer.

Fernanda (her name has been changed to protect the family's identity) was one of the young women in the Ibiuna family present at the first meeting with Achatz and her colleagues round the kitchen table. I met her myself on a visit to the small town, reached just as the sprawl of São Paulo begins to fizzle out into tentative countryside. Blonde, pretty, athletic and tanned, Fernanda sat on the settee with her long legs tucked under her and told me how hard she had been hit by the news that her test result was positive – that she was a carrier of mutant p53 – and how much she dreads the biannual screening trips to A C Camargo. She finds walking through the hospital's bright, crowded cancer clinic on her way to Maria Isabel's consulting rooms, past so many people who are extremely sick, particularly disturbing, she said. That's the fate that awaits herself and her loved ones, she thinks, and memories of her mother's long battle with the disease, and of the frequent visits to hospital, flood her mind.

In Porto Alegre, a man who saw his wife and 11-year-old daughter die from cancer within weeks of each other struggles to cope with the fact that his small son too is a carrier of the faulty gene. He feels the ground has been dug from under his feet and there are no certainties in life any more, said his sister-in-law Margarete; she has taken on responsibility for bringing the child for his check-ups because his father's fears for him are overwhelming. A glamorous, buxom woman in early middle age with large dark eyes, glossy black hair and heavy silver rings on her fingers, she agreed to meet me in a small room off the clinic where Patricia Prolla was seeing patients and their families.

Margarete's face crumpled and her eyes brimmed as she explained that this was her first visit to the hospital since her niece had died of a brain tumour a few months previously. She had sat with the girl for the last 20 days of her life, mostly just holding her hand, as the doctors tried to control the pain. Margarete was thankful she could do something to protect her nephew from cancer, but has not been tested herself for the mutant gene. She shrugged and looked away. She's happier not knowing her status. 'What will be, will be,' she said, drawing irrational comfort from the fact that she looks more like her father, who was not the carrier of the mutant gene, than like her mother, who was.

Until the Brazilian authorities decide how to deal with the public-health crisis posed by the mutant p53, fatalism and grasping at small straws are the only real options for people living with the faulty gene in their families. Meanwhile, LFS is offering insights that might help to resolve one of the longest-running controversies of p53 research – whether or not mutant p53 does indeed act like an oncogene, driving the process of tumour formation under certain circumstances rather than simply knocking out the function of the wild-type allele.

CHAPTER EIGHTEEN

Jekyll and Hyde

In which we hear how researchers came full circle to realise that some mutant versions of p53 do indeed behave like oncogenes, actively driving delinquent cells towards cancer, rather than simply losing their ability to act as tumour suppressors.

***

You say you have to know all these facts – well, clearly the facts, some of them, that you learn are wrong, so if you take them too seriously you won't discover the truth. You could say that if you become too imbued in the ideas, and talk about them too long, maybe your capacity for ever believing they're false would be burned out.

James Watson

When, 10 years after it was discovered, normal p53 was found to be a tumour suppressor not an oncogene (or tumour 'driver'), many people lost interest in the mutants they had been inadvertently studying for so long. Instead, they began to focus their full attention on the wild-type protein, which was a much more exciting prospect. In doing so they chose to ignore what they had observed with the mutants. The few who said hang on a minute, the mutants may indeed be doing something to drive these tumours – something *more* than simply losing their ability to stop the cells running amok – became lone voices talking to empty rooms. 'It was kind of a reaction to the fact that these mutant p53 clones had misled the field and caused us to draw the wrong conclusion,' commented Moshe Oren. 'They were a sore point in our history, and many people just wanted to forget about them.'

Not Varda Rotter. You will remember meeting her back in Chapter 7 in relation to the revolutionary discoveries that led to the recognition of p53 as a tumour suppressor. In 1979–80, her experiments with the Abelson cancer virus had led to malignant blood cells that had no p53 protein at all – a completely different result from that of most of her colleagues, who were finding an over-abundance of p53 protein in their tumour cells. The virus, Rotter discovered, was disabling the gene by inserting a bit of its own genetic material into p53 so that it could not produce any protein.

Curious to see what effect the loss of p53 had on these malignant blood cells, she injected them into laboratory mice, where she found that they caused small tumours to develop that soon regressed. Next she took some of these same malignant blood cells and, using some technical wizardry, replaced the crippled p53 with a functional copy of a mutant p53 gene – that is, a mutant that was able to produce its protein. She then injected these engineered cells into her mice, and this time she found they produced extremely aggressive tumours that were eventually fatal. She published her findings in *Cell* in 1984.

This was dramatic stuff that caught her imagination and when, five years later, normal p53 was revealed as a tumour suppressor, *not* a tumour driver, she was not about to dismiss the aggressive behaviour of the mutant as no longer important. Rotter did not follow the herd nor change the focus of her research and she became the standard bearer of what is known as mutant 'gain of function', often shortened to GOF (remember the analogy in Chapter 7 of the car with the jammed accelerator pedal or the failed brakes? GOF is the jammed accelerator). 'What convinced me was this,' said Rotter, rummaging through her computer images to show me an iconic slide of a thin slice of tumour tissue in which p53 protein showed up as bright red. Clearly the cells were stuffed with it. 'When you take almost any tumour from a human and you stain it for p53, this is what you get . . . Did you ever see anything so covered in protein?' she asked rhetorically. 'I felt it can't just be lying there for nothing.'

That is what prompted her to do the experiments with the mice and the engineered malignant blood cells – she had to prove that her hunch was right and that the abundant protein produced by the mutant p53 gene was actively doing something in tumour cells. This makes p53 unusual among tumour suppressors, almost all of which are simply knocked out by mutation, she explained. 'Unlike other tumour suppressors, p53 has a schizophrenic personality. You have the wild type which is very important: the guardian of the genome that takes care of DNA repair, takes care of genomic fidelity, takes care of everything. But once this is mutated then it becomes a *monster*.' It was this characterisation of p53 that her granddaughter sought to portray in her picture of a devil and an angel that Rotter has framed on her office wall.

Among a field of sceptics, one person who also remained curious about the mutants was Rotter's colleague at the Weizmann Institute, Moshe Oren. It was while he was investigating their activity in dishes in his lab that he stumbled across the temperature-sensitive mutant when the thermostat in one of his incubators started to play up. This, you will recall, led to the discovery that p53 can trigger apoptosis or cell suicide. But it also contributed to another equally critical discovery about the nature of p53 – that the normal, wild-type protein can change its shape, and thereby its behaviour in cells, from a suppressor of growth to a promoter of growth, under certain conditions. Though it's now thought likely to be a common feature of proteins that control many others in a cell, such flexibility of shape and behaviour in a single protein was virtually unknown at the time it was discovered in p53 and, as we'll see, it shed light on mysteries way beyond the field of cancer biology.

AN EXTREMELY FLEXIBLE PROTEIN

The person credited with the discovery of the protein's flexibility is Jo Milner, who developed what is known as 'the conformational hypothesis' of p53. Milner, whom Oren describes as a 'very clever and original' scientist, traces her fascination with biology to a childhood spent in Bridlington, a seaside town on the north-east coast of England, where she spent long happy days walking the beach, guddling in rock pools and coming home with starfish in her pockets. It was her mother who, bringing up her children alone after World War II, nurtured a sense of curiosity and freedom of spirit in Milner and her siblings. Times were tough, she remembered, 'but there was never any sense of hardship. We grew up in a tiny home brimming with friends who all seemed to adopt Mum as their own.

'One of my abiding memories is of looking out of the window from an early-morning train and seeing, across a field, a white sheet being waved from the upper floor of a large house: the train was carrying me to London for an interview at the university; the large house was where my mother worked as a housekeeper; and the sheet was her fond farewell and good luck.'

After gaining a degree in zoology in London, Milner studied for her PhD at Cambridge University. Since then her career has taken her to Harvard, back to England for another 20 years at Cambridge and finally to the University of York, where she was, until very recently, director of the p53 Research Unit in the Department of Biology. I took the train south from my home in Scotland to visit her as she was busy packing up her lab on the brink of retirement and looking forward to following her still-lively curiosity as a scientist without the pressure any longer of leading a team. As we sat in the sun room of her elegant stone house in an old village nestled in farmland outside York, the scent from an ornamental lime tree in a pot hanging delicately in the air, she talked of the steps that led to her discovery of p53's extraordinary flexibility – one of the rare and thrilling eureka moments in a scientist's life.

A question that intrigued Milner about biology in general

was how cells switch from a dormant, or quiescent, state to a dividing state as tissues grow or repair themselves in the normal course of life. She was investigating this question in healthy white blood cells, which remain happily quiescent in their nutrient-rich culture medium in the lab until stimulated to enter the cycle of division. For her experiments she was using antibiotic drugs as biochemical 'tools'. One of these antibiotics was a toxin derived from the death-cap mushroom that works by inhibiting the synthesis of essential proteins and cutting off the metabolism in cells so that they grind to a halt and die.

Milner found, however, that if she exposed the cells in her culture only briefly to the toxin, the effect was reversible. She developed a method that enabled her to put the brakes on in her cells just at the moment they were about to enter division and then, by removing the toxin, to release the brakes and let the cells carry on through the cycle. Using this method, she found clear evidence of a gene expressing a protein that appeared briefly in her cells and stimulated them to switch from quiescence to division, before it was degraded and disappeared. 'Obviously the next question was what that gene might be, and what was the protein involved,' she commented.

p53 had recently been discovered and, after reading the papers by Lane and Levine, Milner thought it would be interesting, just on the off chance, to check this gene out. She managed to get hold of two different antibodies that had been tailor-made to recognise and flag up p53 when it was present in cells. Using these antibodies she found that p53 was indeed expressed, and that one of the antibodies recognised the protein in the quiescent cell and the other recognised it as the cell began to divide. Neither antibody recognised the protein in both states. Here, it seemed, was a good candidate for the genetic switch. But why was the protein different – so different that it was undetectable by some tailor-made antibodies to p53 – during each stage of the cell cycle? 'The only thing I could think of,' said Milner, 'was that here was a protein that was changing conformation (or structure). So you have one epitope (the face of a molecule to which an antibody attaches) exposed and another hidden away in one conformation.' She clenched her fist to demonstrate a folded protein showing its face to the antibody. 'And then when you stimulate it, the conformation changes and the other epitope is exposed.' She opened her fist slightly to show another face as the protein refolded itself.

This was a revolutionary idea, and her results raised the question of whether this was one protein changing shape or two slightly different versions of the protein produced by the same gene in order to flip the switch in the cell. It's here that Moshe Oren's temperature-sensitive mutant comes back into the picture. Just as Milner and her team were asking themselves about the nature of the p53 protein that was throwing the switch in dividing cells, she and her lab technician attended a p53 conference at which they heard Oren speak of his serendipitous findings, and a light went on in Milner's head. 'I thought: fantastic! We can check his mutant to see if our conformation idea is right,' she recalled.

Her idea was to look at how a single blob of protein produced by the mutant p53 folded itself at the different temperatures, to see if this was what dictated its changes in behaviour from a growth suppressor to a growth promoter and back. When she suggested such an experiment to Oren he was doubtful it would work, fearing that the techniques involved, which included a period of incubation on ice, would upset the mutant's temperature sensitivity. He had not considered the protein's conformation, and anyway he had other research ideas in mind for his mutant.

However, Milner still believed the experiment was worth a try and, together with her technician, went ahead as soon as she arrived back in Cambridge. 'It was just beautiful!' she said with a big smile at the memory. 'That was a real high point, because for a long, long time we'd been trying to get a handle on conformation, juggling with different conditions to see if we could induce any change. And here we had this mutant that did it perfectly. It was one of the special moments in life . . .' Its great significance was the revelation that p53 can switch behaviour – from a suppressor to a promoter of growth and back again – without the need for mutation. *Both* roles, it seems, are part of the protein's normal repertoire, and this flexibility of form is what makes it possible. Its flexibility is also an explanation for how p53 is able to play such a varied, subtle and central role in cells, both normal and diseased. It has turned out that the concept is much broader than just p53, said Oren. 'But p53, in my mind, set the paradigm for this duality of function.'

THE 'DOMINANT-NEGATIVE' EFFECT

The discovery of p53's intrinsic shape-shifting nature opened the door to all kinds of experiments designed to help us understand the activity of the gene in more detail. Milner's next step was to investigate a phenomenon known as the 'dominant-negative' effect, in which the behaviour of cells that have a wild type and a mutant copy of the same gene, both active, is dominated by the mutant. This had never been seen with a tumour-suppressor gene: all those discovered thus far conformed to Knudson's model in which a cell that still has one wild-type allele will function as normal until that allele is knocked out by some event. In other words, the wild type dominates over a mutant.

However, many of the people who remained sceptical of the gain-of-function theory – that mutant p53 produces a protein with new and abnormal functions – suggested that the dominant-negative effect might well be what people had observed with p53 and mistaken for gain of function. After all, they were used to this tumour suppressor breaking the rules. But if this were the case, it was not true gain of function people were seeing at all, but *loss* of function by an unusual route – by the wild-type p53 being, as it were, overwhelmed by the mutant which crippled its brakes.

No one knew how this might occur, however, and the object of Milner's next experiments was to try to understand the relationship between different versions of p53 operating in the same cell. Do they work as separate agents, or stick together to form a co-operative unit? To make things clear, she used combinations of mouse and human p53 in her test tube because, being slightly different-sized molecules, the two proteins could be easily followed in her experiments. As with two different-coloured but similar blocks of Lego, you could see how they fitted together, if they did, rather than being confronted with an amorphous blob.

Milner already knew that p53 protein molecules can clump together in groups of two or four to form co-operative units. Now she showed that this assembly was restricted to proteins of the same conformation. Thus she found that mixing the pre-formed proteins in her test tubes gave either suppressor-suppressor complexes or promoter-promoter complexes, but not suppressor-promoter complexes. Clearly, the affinity between the p53 building blocks was determined by their shape.

However, in real life the dominant-negative effect occurs when wild-type and mutant proteins are co-expressed – that is, produced together and simultaneously in the same cell by the two different alleles, or copies, of a single gene. So Milner and her colleague simulated this scenario in the lab, co-expressing the suppressor (wild-type) and promoter (mutant) forms of p53 side by side and simultaneously in the same mixture. This was the acid test, and the results were spectacular: not only did the two co-expressed proteins form a complex, but the only antibody that recognised the new unit was the one tailored to the promoter (mutant) form. The dominant-negative effect people had speculated about was a real possibility, and here was a novel mechanism to explain it – a clear demonstration that one misfolded p53 protein in a co-operative unit of proteins can force the others to change shape in domino-like fashion.

Milner's research, conducted under artificial conditions in the lab, was proof of principle; no one knew if this is what happens in real life. But her findings, published in *Cell* in 1991, quickly caught the attention of Stanley Prusiner, a scientist with an equally original mind working in a very different field – that of the so-called 'spongiform encephalopathies' that include mad cow disease and its human equivalent, Creutzfeldt-Jakob disease or CJD, as well as scrapie in sheep. His ideas about how these diseases might arise had been widely scorned as heretical and he was looking for just such a mechanism as Milner described to strengthen his case.

In 1972, Prusiner, then working as a neurologist at the University of California, San Francisco, had admitted a female patient to his ward suffering from CJD, which kills nerve cells in the brain, leaving holes that give it the characteristic sponge-like texture. His patient was progressively losing her memory and her ability to perform routine tasks, and Prusiner was told she was dying of a 'slow virus' infection. However, in years of research, no one had been able to pin down this slow virus, so-called because of the long incubation period between exposure to the agent and the appearance of symptoms.

'The amazing properties of the presumed causative "slow virus" captivated my imagination, and I began to think that defining the molecular structure of this elusive agent might be a wonderful research project,' Prusiner wrote some years later in an autobiographical sketch. His efforts gradually convinced him that he was dealing not with a virus, nor with any other known infectious agent such as a bacterium or a fungus, but with a misfolded protein – and he named his novel pathogen a 'prion'. But how could a substance with no DNA to carry the instructions of replication transmit a disease? This was the heresy that caused the firestorm when Prusiner published his prion hypothesis in 1982. 'Virologists were generally incredulous and some investigators working on scrapie and CJD were irate,' he wrote. 'The term prion, derived from "protein" and "infectious", provided a challenge to find the nucleic acid of the putative "scrapie virus". Should such a nucleic acid be found, then the word prion would disappear!'

Of course no DNA was ever found, and as evidence mounted for his novel theory of infection, the sometimes vicious personal attacks from his critics gradually died down. But still Prusiner needed an explanation of how a misfolded protein might corrupt the normal protein we all have in our brains. 'Our paper was the very first evidence that such a thing could happen,' commented Milner. 'Prusiner was visiting a colleague in Germany at the time and contacted me to arrange a meeting. We met in my office in Cambridge and talked for three hours before I drove him back to the station. It was so exciting to exchange ideas – just lovely!' Six years later, in 1997, Prusiner was awarded the Nobel Prize for Medicine for his prion hypothesis which, though it still has its critics, is now widely accepted as the explanation for the deadly spongiform brain diseases.

A LIFE OF ITS OWN?

Not all p53 mutations produce so-called 'conformational' (or 'structural') mutants that behave this way. 'Contact' mutants, which produce a protein unable to attach to DNA and switch on other genes, are the type most commonly found in human tumours; in these cases the wild-type protein will win the day, keeping the mutant in check until that good copy of the gene is lost in the course of living. As the unusual variability of p53 mutants became apparent, the debate about 'loss of function' versus 'gain of function' became ever more intense and, in the late 1990s and early 2000s, a number of research groups created transgenic mice to try to resolve it and to find out what happens in real life.

Guillermina ('Gigi') Lozano, whom we met working with mouse models at MD Anderson in Houston in Chapter 13, headed one such group. Lozano's family had immigrated to the US from Mexico in search of a better life, and Gigi was the first among them to go to college. In 1986 she earned a doctorate in biochemistry from Rutgers University in New Jersey, but as a postdoc she chose to join Arnie Levine's molecular biology lab at Princeton, attracted by the fact that he was working on one of the very first mouse-tumour models. 'When I realised you can manipulate the mouse genome to mimic the kinds of tumours you find in human cancer, I was fascinated. There was no going back for me,' she told me with a grin when I met her at a mutant p53 conference in Toronto in 2013.

Trained by Levine, Lozano got a job in molecular genetics at MD Anderson, where she is now Professor and Head of the Department of Cancer Genetics. Much of her research involves mouse models and in the early 2000s she set about creating one that mimics the human Li-Fraumeni syndrome, in which the p53 gene has one wild-type allele and one allele with a 'point' mutation, meaning that just a single letter in its code is changed. People had become adept at creating knock-out mice, with a whole gene or one of the two alleles 'deleted' from the DNA, but a knock-in mouse – one with a point mutation – was an altogether trickier proposition that took time, skill and patience. Lozano and her group chose a mutation that corresponds to the R175H mutation in human cancers 'because it's the worst mutant you can possibly have,' she explained. 'And if you're going to generate a mouse for the first time you don't want a mutant that's kind of wimpy.'

Meanwhile, at MIT in Boston, Tyler Jacks – renowned for creating one of the two first p53 knock-out mice in 1992 – was on the same track. His lab was busy generating two different mouse models that mimicked LFS – one with the same point mutation as Lozano's mice, corresponding to human R175H, and another corresponding to R273H. The two groups published their findings in the same edition of *Cell* in December 2004. What distinguished their mouse models from others designed to test the activity of mutant p53 was that here the gene was being switched on *naturally* in response to signals from the cell's environment. In most other models, the gene was switched on artificially by the researchers – and herein lay the big sticking point. The sceptics argued that in all experiments that appeared to show gain of function, the gene had been over-stimulated by the researchers, leading to an over-abundance of the protein. The artificial manipulation by researchers was bound to upset the delicate machinery of the cell, they said, and they were not convinced this pooling of protein – and hence gain of function – is what happens in real life.

However, both Lozano's and Jacks' mice showed that indeed the mutant protein does pool in real life. They furnished compelling evidence, too, that gain of function – by some p53 mutants at least – is a real phenomenon and not an artefact produced by force. The two groups used different strains of mice in their experiments, which meant that the mutant genes were operating against a variety of background environments, thus adding weight to their findings. As controls for their experiments, both groups used a mouse with one wild-type allele and the other p53 allele missing altogether; these control mice, too, were prone to cancer, but mutation would be no part of the picture, allowing the researchers to see, by comparison with their LFS-like mice, what effect, if any, mutation had on the types of tumours that developed.

So what did they find? As one would expect with living creatures, the different mice met different fates. They all – those with one mutant p53 allele (the LFS-like mice) and those with one missing allele (the controls) – developed tumours. In all Jacks' mice with mutant p53 the range of tumours that developed was different from that seen in the control mice. In Lozano's R175H mutants, on the other hand, the tumours that developed were similar to those of their controls, but they were much more aggressive: they spread readily to the lymph nodes, lungs, liver and brains of the mice, while the tumours of the controls did not metastasise.

'To me that was the most convincing experiment,' Lozano commented. 'When you compare those two mice and one has a tendency to produce tumours that metastasise and the other one doesn't, how can you argue against gain of function? I mean, you can't!' What finally clinched the argument for most of the p53 community – including some die-hard sceptics of gain of function – was that both Lozano's and Jacks' mice developed some novel tumours that are never seen in knock-out mice with *no* p53. This could mean only one thing: that the mutant was doing more than simply hobble the wild-type allele and shut down its protective functions – clearly it had a life of its own.

These and other mouse models have allowed researchers gradually to build a picture of how the mutants work and how they interact with wild-type p53. Context, it seems, is all-important. Not only do the mutants differ from one another in their actions, but they behave differently in one cell type, tissue or organ from another, and in one strain of mouse (and presumably one human being) from another. Timing, too, is critical: in some cancers p53 mutation is an early event; in others it occurs when the tumour is already advanced, and may (as in the case of colon cancer) mark the turning point between a benign growth and malignancy.

As for their mechanism of action, it seems that p53 mutants sometimes co-operate with other oncogenes, such as Ras, to drive the growth of tumours. Sometimes they achieve the same effect through interaction with another protein in the cell – notably one or other of p53's closest relatives, p63 or p73, which share some of its tumour-suppressor characteristics and can be hobbled by the mutant. Some mutants can, like wild-type p53, switch on and orchestrate the activity of other genes. However, this is a travesty of healthy behaviour: the genes switched on by mutant p53 are not the same genes that are controlled by the wild-type tumour suppressor, and can have the opposite effect. And a distinctive characteristic of many of the mutants is that they make cells extremely resistant to self-destruct signals. This not only encourages the growth of tumours, but makes them very difficult to treat, since most anti-cancer therapies are designed to trigger the apoptosis response.

An unexpected finding, made by Gigi Lozano and her group, is that the over-expression of mutant p53 protein is not an intrinsic property of the mutant gene, as had been assumed. The pooling of the protein occurs only in tumour cells, while in the normal cells surrounding the tumour and beyond, the protein is at barely detectable levels. This implies that, like the wild-type protein, the mutant is being regularly produced and degraded in the normal course of events until something occurs to release it from the loop. Though theories abound, no one yet knows why or how this happens – only that the mutant protein *has* to be over-expressed to be able to act as a growth promoter.

VINDICATION

As Varda Rotter's steadfast insistence on the importance of mutant p53 has been vindicated and the spotlight has swung back in this direction, enormous effort is being made to understand its biology. 'Over the last five years alone,' wrote Carol Prives and William Freed-Pastor in a review for Cold Spring Harbor Laboratory Press in 2012, 'p53 mutants have been found to actively contribute to tumor proliferation, survival, limitless replication, somatic cell reprogramming (i.e. turning differentiated body cells back towards stem cells), genomic instability, inflammation, disruption of tissue architecture, migration, invasion, angiogenesis (development of a blood supply to a tumour), and metastasis.' They concluded that this confirms mutant p53's central role in the development of malignant tumours, with an impact on nearly all of the 'hallmarks of cancer'– the list of 10 defining characteristics of all cancers – proposed by Bob Weinberg and Doug Hanahan in 2000.

It also makes the aberrant protein a prime target for therapy, as scientists and Big Pharma alike look for new, more effective ways to treat people with cancer that do not do such devastating damage to the body's normal, fast-dividing cells at the same time.

Cancer and Ageing

CHAPTER NINETEEN

Cancer and Ageing: a Balancing Act?

In which we learn that ageing is the price we pay for protection from cancer: wrinkles, sagging tissues and thinning bones are the result of cell senescence and gradual depletion of stem cells, the body's repair materials, through apoptosis.

***

Research at its best is the finding of answers to questions about the world that have not previously been asked.

John Maddox

As a mouse man, Larry Donehower of Baylor College of Medicine in Houston is used to dropping bombshells. Working closely with his colleague Allan Bradley, he was, in 1992, the first person to create a p53 knock-out mouse. Using the technique recently developed by Capecchi, Evans and Smithies, he was able to delete the gene from mouse embryonic stem cells and then implant the developing embryos successfully into the womb of a female mouse for gestation. When he turned up to present his findings at a p53 meeting that year, the excitement was palpable and most people had a pretty good idea of what he would say. After all, it was not long since the normal gene had been revealed as a tumour suppressor, not an oncogene; it had been found in almost every multi-cellular organism, conserved in evolution, unchanged in form and function, since the dawn of time; and it had recently been dubbed 'guardian of the genome'. Surely Donehower's mice would show that without it, life was not sustainable?

But this was not what he had come to say. His genetically

engineered animals were fine. Not only had they survived without the protection of the guardian the period of explosive growth, cell division and differentiation that turns an embryo into a pup, but they had no signs of physical deformities or cancerous growth. Donehower's audience was stunned.

David Lane remembered the occasion vividly when I interviewed him for this book on the fringes of a big conference in Liverpool. 'We were all in a very triumphant mood as a community. "p53 is now the most important bloody protein in the world and all you guys can get stuffed!", you know? "Who's been telling us that we've been wasting our time for the last 10 years?" It felt exciting and good,' he recalled with a grin. 'We were having this big p53 workshop in the US, and instead of just 20 people coming, *200* were coming. There were lots of positive data . . . loads of people finding mutations, and everything was now very convincing. Then Larry stands up and says, "I've made a knock-out mouse and there are some interesting things about this mouse . . . First of all it's completely viable. There are no defects that I can see – and it certainly hasn't got any cancer!" Everybody went, "Uh oh!"' Lane leaned back in his chair and gave a huge laugh of incredulity. 'Of course, I guess it was about two months after the conference, Larry started to see massive development of tumours in all the animals and there was a collective sigh of relief . . . Poor mice, but lucky us!' Lane paused to reflect over the 20 years since that meeting. 'It was incredible, actually. And, of course, having the knock-out as a tool has made an enormous difference to everything.'

Donehower had been as stunned as everyone else by his initial results, which had ramifications beyond the purely scientific. The colossal effort that goes into such research and the money poured in to support it generate high expectations and heavy pressure for exciting results. When 'nothing' happens, panic sets in and doubt ripples across the whole community. In time, however, all Donehower's knock-out mice did indeed succumb to cancer, surviving for less than five months compared with around 30 months for normal mice of the same genetic background. Further research also revealed that p53 knock-out mice have much smaller litters than normal, suggesting the gene plays a part somewhere in reproduction. And it has flagged up a vital role for p53 in regulating metabolism, which is one of the hottest topics of investigation at the moment.

But back to the story of ageing. In 2002, Larry Donehower and his team dropped their second bombshell when they made a mistake with one of their experiments and got a mighty surprise. They were trying to make a knock-out mouse using a different technique from before, but they ended up instead with a mouse in which the still-present p53 gene was hyperactive. Sure enough, the creatures proved well protected from cancer, as the researchers would have predicted. What none of them expected to see, however, was that they aged exceptionally fast. In just a few months, they looked like very old mice. 'They had hunchback spines, ruffled fur, grey hair; things like that. And they lived only about two-thirds of their normal life span,' Donehower told me when I spoke to him at a p53 meeting in New York. 'Some of the most interesting findings in science are accidental, actually. They're not what you're looking for or expecting, and this was very surprising. *Nature* published it in 2002. Now this accidental finding is opening up a whole new area of research about how this very important cancer gene can also modify the ageing process.'

People have known for a long time that ageing and cancer are related, in that the chances of getting cancer increase with age. But not even the scientists suspected they might be two sides of the same coin, sharing a common mechanism in which the scales can be tipped either way. In other words, that wrinkled skin, thinning bones and failing organs may be the price we pay in the long run for holding cancer at bay. Donehower's findings, however, left room for a smidgen of doubt about the role of p53, since the 'accident' that produced the hyperactive version also knocked out a stretch of DNA upstream of the tumour-suppressor gene. The possibility could not be ruled out that something here might be responsible for the premature ageing. But soon another lab, run by Heidi Scrable of the University of Virginia at Charlottesville, provided new evidence that Donehower's original hunch was right. She and her team created a mouse model in which the only change to its DNA was the replacement of one allele of p53 with a naturally occurring hyperactive version of the gene, and found the same thing – premature ageing and death.

Donehower, Scrable and others working in this compulsively intriguing field have gradually pieced together the picture of how this can happen. A hormone known as insulin-like growth factor 1 (most often represented as IGF-1) that, not surprisingly, plays a central role in the growth and proliferation of cells, has long been known to promote ageing too in all manner of organisms, from fruit flies and nematode worms to mice. By tinkering with the strength of the signals this hormone sends out to the cells, researchers have managed to manipulate the life span of these creatures. The effect is most obvious in flies and worms, which live considerably longer when IGF-1 signalling is dampened down and shorter when it is amplified.

Scrable and her team found that the hyperactive p53 in their engineered mice stimulated hyperactivity of the growth hormone too. The amplified signals from IGF-1 in turn triggered the mechanism designed to bring runaway cells under control by driving them into senescence, or irreversible arrest. This, of course, is tumour suppression at work, and is orchestrated by 'regular' p53. To that extent it was an appropriate, and clearly beneficial, response. But senescent cells can become dysfunctional and as they accumulate in the tissues they begin to cause trouble of their own.

Unlike cells that have been driven to suicide by apoptosis, senescent cells remain alive and active – and, significantly, they alter the micro-environment of the tissues by secreting proteins that communicate with neighbouring cells and even with distant organs. Some of these proteins are important for tumour suppression – for example, they inhibit the development of new blood vessels which might feed a developing tumour. But as they metabolise in the normal course of events, senescent cells also produce large amounts of material that seeps into the surrounding tissue. 'This begins to chew up the extra-cellular matrix – you know, the stuff that keeps cells glued together,' said Judith Campisi, who studies senescence at the Buck Institute for Research on Aging in Berkeley, California, when I spoke to her at the same New York meeting as Larry Donehower. 'The major extra-cellular molecule that keeps your skin looking young is collagen. And sure enough, senescent cells produce molecules that destroy collagen.' Hence wrinkles.

Campisi, dark haired and petite, with dangly earrings and the graceful posture of a ballet dancer, uses her hands and eyes expressively as she speaks. She began her research career focusing on cancer, and it was here she first encountered senescent cells, in the context of tumour suppression. But she soon became fascinated by their possible role in the normal processes of ageing – an idea that all but a small 'crazy contingent' of scientists had dismissed for a long time. 'I didn't buy [the theory] for a minute,' she said in an interview with one of her colleagues at the Buck Institute in early 2013. '[But] sure enough, we started working on this problem and I had to realise that this "crazy contingent" was actually correct!'

To the uninitiated, the scientists' original doubts seem strange, since the very term 'senescence' implies ageing. This was clearly intended by Leonard Hayflick, the man who discovered and named the cells in 1961 – but he was putting himself out on a fragile limb scientifically, Campisi told me. 'Len Hayflick was a cell biologist, and he was studying cell proliferation for a very specific reason: he was interested in growing viruses in human cell cultures as opposed to animal cell cultures, and virologists were having a helluva time. They would get these cultures and initially they'd do great, and then eventually they wouldn't do so great, and they'd throw them out and start again.' Hayflick decided to study this in much more detail, and he made the startling discovery that, in contrast to most cancer cells, normal human cells have a finite ability to undergo cell division in culture. 'Now that finite ability is huge,' said Campisi. 'For stem cells taken from a human embryo we're talking 40, 50, 60 population doublings. So you can see why you'd be fooled – you'd do an experiment for three months, the cells are growing great and then they start not to do so great and then they stop.

'Hayflick made two interesting observations,' she continued. 'One of them was obvious and was immediately accepted, and that is, "My God, tumour cells don't do this. Maybe this is a way of stopping cancer." It fuelled a whole area of research to think about the senescence process, which is controlled by this famous suppressor gene, p53, as stopping cancer, and I think there's very little controversy about that now. But he also made another observation that was totally unscientific, totally intuitive, based only on his sense as a cell biologist. He looked in the microscope; he looked at these cells and said, "They look old." Now what on earth does that mean, that a cell looks "old"? I mean it's an unquantifiable observation. But he said, "Maybe what's also happening is that we are recapitulating some aspects of ageing in a culture dish." That observation, that comment, went largely unnoticed except for a few, again pretty imaginative, people in the field who picked this up and began to study senescence, not as a tumour-suppressive mechanism but as an ageing process. It is still somewhat controversial – less so than it was 50 years ago, but still controversial, though it's also gained a lot of momentum.'

The number of divisions a normal cell can undergo before becoming senescent is known today as the Hayflick Limit, and it is measured by the telomeres on the ends of the chromosomes. Telomeres are protective tips to the chromosomes that are rather like the little plastic caps put on the ends of shoelaces to stop them unravelling. Every time a cell divides, the telomeres shorten, until they are no longer able to protect the chromosomes and the cell goes into permanent arrest. And though this is not the only route to cell senescence, dangerously shortened telomeres are one of the stressors that trigger the p53 response.

Today Judith Campisi is a world leader in the field of cell senescence, and she is in no doubt that these cells are at the pivotal point of a mechanism that can tip either way. 'What my work tries to do is to reconcile the two very different views of senescence. One says it's really good for you, it stops cancer. The other says it happens during ageing and it looks like it's bad for you because the cells look kind of old and ragged.'

Her lab has discovered recently that senescent cells provoke inflammation – a condition that underlies almost every major age-related disease, she told her interviewer at the Buck Institute. 'We've shown now very clearly that one senescent cell sitting in a sea of non-senescent cells will provoke an inflammation that will spread to other cells. So it's a very appealing hypothesis that you don't need very many senescent cells to be able to drive the degenerative changes that are a characteristic of ageing organisms.'

Senescent cells are also highly resistant to apoptosis and the ultimate irony is that, with time, they themselves become a cancer risk, helping to drive the process of uncontrolled growth. But how? 'In the last couple of years we've learnt that the senescence response has another life, and that is to promote tissue repair when needed,' said Campisi. 'That's where it begins to be problematic in later life.' As senescent cells become dysfunctional with time, she explained, they can start to send out signals to initiate tissue repair and proliferation of cells in the absence of real injury, thus driving the development of tumours.

This, of course, is not inevitable. As researchers have found with every aspect of tumour suppression, context is all important: different cell types and tissues follow different paths on activation of p53. Scott Lowe, another mouse-model man, whom we met in the chapter on apoptosis, is also at the cutting edge of cell-senescence research; he discovered that, although these cells are indeed resistant to killing by apoptosis, they don't always hang around in the tissue to become toxic. In some tissues they communicate with the immune system, which sends in the scavenger cells to clear them away.

'We had this study in 2007, in which if you had no p53 you had a cancer cell; if you flipped p53 on, the cancer cell went senescent,' Lowe explained. 'We could see this in the Petri dish and also in the animal. But whereas in the Petri dish the cells just sat there – they never divided, they didn't die but they didn't grow – in the animal the tumour went away.' This was perplexing: surely, the researchers figured, if the cells weren't dividing, nor were they dying, the tumour should stay the same size. So what was happening here? Digging more deeply into the mechanism, they found that proteins secreted by the senescent cells were triggering an immune response that was removing them as effectively as apoptosis. What's more, they could watch this happening in the lab, watch the scavenger cells engulf senescent cells, if they put the two together in the Petri dish and allowed them to communicate – in effect creating a simplified version of the community of cells you would find in a living body.

As with their discovery of dead cells and apoptosis, this was a surprising result, not what they were expecting to see at all, but Lowe found it strangely pleasing intellectually. 'For 15, 20 years, we studied how p53 affects the cell that it's turned on in, but now we realised it does more than that . . . It also can send signals that affect the surrounding tissue.'

Still today no one knows exactly why the stress response leads to different outcomes under different circumstances. 'Part of it is tissue-dependent: lymphoid cells will, by and large, apoptose, and connective tissue will senesce,' said Lowe. 'But it isn't completely that. There are other factors that influence it, some of which we know, but none in a way that you can satisfyingly say is decisively the answer . . . And it's not even that p53 is turning on different sets of genes when the cells die versus when they arrest, so it's also something about how the cell *interprets* the genes that p53 turns on. This is a really interesting question for what we now call "systems biology" – how the cell integrates multiple signals to make a yes or no decision to go down a certain path – and it's stuff we study to this day.'

Researchers interested in the gene's role in ageing believe that both apoptosis and senescence are significant to the process – senescence for all the reasons discussed above, and apoptosis because it gradually depletes the pool of stem cells our bodies need for repair and maintenance. 'The simplest model would be that you're born with a limited number of stem cells,' explained David Lane. 'Those stem cells are very easily killed off by DNA damage, so they're the ones most tightly controlled by p53. If you set a stress-response threshold where they're too easily killed, then you don't get cancer but you run out of stem cells more quickly. If you set the threshold such that they're hard to kill, then you could live a long time, but you're more likely to get cancer.'

Age researchers also have a theory, drawn from evolutionary biology, to explain the paradox of why a system designed to preserve life by protecting us from cancer should also drive the mechanism that leads inexorably to our decline. Basically, nature only cares about perpetuating the species, so evolutionary pressures to select for advantageous traits – and weed out ones that are harmful – operate only up to and across our reproductive years. Beyond that we are living on borrowed time: nature no longer has a use for us, and natural selection is a spent force.

Ageing is anyway a modern phenomenon. During the great majority of our time on Earth, humans didn't die of ageing: we didn't die of cancer or Alzheimer's disease, or even cardiovascular disease; we died of accidents and predators and infection and starvation. Thus for millennia, ageing operated below the radar of evolution by natural selection. It's only today, as we've conquered infection, hunger and predation in much of the world, that the damaging flip-side of tumour suppression has been able to play itself out to full effect. But understanding the roots of ageing holds promise for the future. 'People are beginning to ask: can I manipulate the system to get the best of both worlds?' commented David Lane. 'Can I sensitise the gene for short periods (to eliminate cancerous cells) and can I suppress it (to keep ageing at bay)? I think one can imagine really quite extraordinary results as we begin to be able to control this system.'

CHAPTER TWENTY

The Treatment Revolution

In which we hear of p53's place at the cutting edge of gene therapy and personalised medicine, which are revolutionising the treatment of cancer – and, some predict, will remove the threat of ever dying of cancer from today's young people.

***

[p53] *already, with no help from doctors, stops incipient cancer millions of times every day. Scientists do not have to top the elegant system that nature has engineered. They just have to harness it.*

Sharon Begley

'If you peer into cancer cells – and we've got amazing technologies now to catalogue all the things that are happening – and you look at them from an evolutionary point of view, it turns out that even within a single tumour there are many, many different subspecies of cells that have evolved in slightly different ways,' says Gerard Evan, whose remarks about the rarity of cancer opened this book. 'And here we are, basically trying to wipe out the entire tumour while keeping the patient alive at the same time. It's a tough deal!'

But Evan is not disheartened. Indeed, as evidence mounts that cancer is an even more complex disease than anyone realised – a hotbed of evolution that makes of tumours a constantly moving target for therapy – he remains decidedly optimistic. Why? 'Let me give you an analogy,' he says. 'Absent of cancer, human beings have been subject to terrible diseases of cells that grow inside them and invade and spread. These cells are genetically very heterogeneous, they exchange genetic information one with the other, and they grow like crazy . . . They're called bacteria, right?

'A hundred years ago you'd have looked at all the

infectious diseases and said, "Oh my God, we've got to have a cure for each one – there's TB in the lung, and in the bone, and then there's this and that . . ." But it turns out they share a great deal of commonality, and if we hit them with antibiotics we can more or less eradicate, at least for a time, infectious disease due to bacteria. Now the fact is, bacteria are much, *much* more genetically complex and heterogeneous and hardy and resourceful in the evolutionary sense than cancer cells.'

The task for drug developers, Evan believes, is to find the commonality of cancer – the 'mission critical' mutation without which no tumour can survive. Not everyone agrees with this analysis; most cancer researchers are still backing the idea of targeted therapy tailored to the individual patient's tumour characteristics. But whatever the perspective, tumour suppressors are obvious candidates for investigation, and much of the effort of academic researchers and their counterparts in the pharmaceutical industry is focused on repairing or enhancing the body's natural capacity to single out and eliminate rogue cells.

People have high hopes and many imaginative ideas for p53-based therapies, though the journey from the lab to the patient's bedside is often frustratingly slow. Ironically, as the explosive speed of technological advance makes it ever easier and quicker for scientists to develop potential new drugs, the rules and regulations governing the process get ever more tight: it typically takes a decade or more for a promising new therapy to be approved for use on patients. Very many prototypes never make it that far; drug development is, by its very nature, a painstaking process of trial and error, but even the 'failures' teach valuable lessons along the way.

VIRUSES AS DRUGS

The first person to try p53-based therapy in humans was Jack Roth, who in 1996 recruited to his study nine patients with inoperable lung cancer whose tumours were no longer responding to conventional therapy. In a pleasing twist to the story, Roth's therapy made a virtue of the pernicious properties of viruses – the fact that the only way they can survive and reproduce is to invade the living cells of the host organism and hijack the machinery of replication. Using genetic engineering, he and his colleagues converted a common virus into a vehicle for transporting good copies of p53 into cells where the gene is dysfunctional. This engineered virus they injected direct into the patients' tumours and found, to their gratification, that the strategy worked: the p53 gene was successfully transferred to the tumour cells; it switched on to produce healthy protein, and the patients suffered no significant side effects.

However, the viral vector, or delivery vehicle, proved poor at evading the sentries of the immune system, and in subsequent prototypes the scientists coated the virus with a substance to give it a better chance of killing tumour cells before being wiped out itself by the immune system. They also changed the delivery vehicle from a retrovirus to an adenovirus – the type that causes the common cold and other respiratory infections.

Therapies based on this design have been tested now in thousands of patients in clinical trials mostly in the US and China. They have proved effective, especially when used in conjunction with conventional chemo- or radiotherapy. Patients also need to be carefully selected for their suitability, since the treatment works better under some conditions than others. It tends to be most effective, for example, in tumour cells in which existing wild-type p53 protein is trapped by over-expression of its natural controller Mdm2; or when mutant p53 protein is produced at such low levels in the cancer cells that it cannot overwhelm the wild-type protein produced by the gene therapy. (You will remember that in some cases where a person has a mutant and a wild-type copy of the p53 gene, the mutant protein is powerful enough to knock out the function of the wild-type protein – the so-called 'dominant-negative' effect. Someone with such a powerful mutant will not be a good candidate for the gene-transfer therapy.)

Roth's pioneering work in the mid-1990s led to the development of two trademarked products, Advexin in the US and Gendicine in China. As well as being injected directly into the tumour, these can be administered by injection into an artery or vein like chemotherapy, and have been tested in a number of tumour types including lung, liver, and head and neck with varying degrees of success. They seem to be effective also, used alone, in preventing early lesions in the mouth from turning malignant. In 2003, Gendicine was approved by the Chinese regulatory authorities for use in the clinic. The first gene-therapy product to receive official approval anywhere in the world, it is used today, in conjunction with radiation, to treat patients with head and neck cancer in China.

In 2007, Zhang Shanwen of Beijing Cancer Hospital, who chaired the clinical trials of Gendicine, gave an indication of its effectiveness. At a conference in China, he presented data from a trial in which 26 patients were treated with gene therapy plus radiation and 27 controls were given radiotherapy alone. Seventeen of the 26 patients who received the combined therapy were still alive five years later, of whom 16 remained completely tumour-free. Of the 27 controls given radiotherapy alone, 14 were still alive five years later, 10 of them tumour-free.

However, despite some remarkable individual success stories and despite being almost identical to Gendicine, the US product, Advexin, has had a rocky ride. The quest to get this product – considered an 'orphan drug' because of its limited market potential as a therapy primarily for head and neck cancer – into the clinic has been enormously expensive. When the US Food and Drug Administration (FDA) declined approval in September 2008 because there was not enough evidence of its effectiveness, the manufacturer, Introgen Therapeutics Inc. of Houston, went bankrupt. Today, Vivante, a small company that rose from the ashes of Introgen and was itself acquired in 2010 by the Swiss-based giant Lonza, holds the licence for Advexin and continues the quest for approval from regulatory authorities in the US and Europe. Meanwhile, the Chinese manufacturer of Gendicine is also seeking FDA approval for its product in the US and in India.

In the early 1990s, Frank McCormick at the University of California, San Francisco, began developing a therapy that uses the common-cold virus in a very different way. He had observed that cancer cells and adenoviruses share some important characteristics, one of which is that in order to stay alive they need p53 to be out of action. Here was a trait he could exploit. But it needed a good deal of engineering to ensure that the virus would target and kill *only* cancer cells and not cause more widespread infection. Essentially, McCormick removed the mechanism by which the virus itself normally knocks out p53 when it enters our bodies. This meant that his engineered virus could survive only in cells which already had no functioning p53 – that is, cancer cells. In these the virus grows and multiplies until the cells literally burst. However, if the engineered virus invades cells with functioning p53 – i.e. non-malignant cells – it withers and dies because it no longer has the machinery to knock out the tumour suppressor. The process by which the cancer cells burst is known as oncolysis, and part of the beauty of McCormick's mechanism as a therapy is that engineered virus particles spilling from the burst cells can infect and destroy neighbouring cancer cells in the same way, but pose no threat to normal cells in the body.

In 1992 McCormick co-founded Onyx Pharmaceuticals Inc. to develop his idea, and in 1996 the therapeutic agent ONYX-015 entered clinical trials in the US – the first engineered oncolytic virus ever to be tested in humans. Those early trials, first on patients with head and neck cancer and then on those with a variety of other tumour types, looked good. The gene-therapy community was riding high. Then came a body blow.

In 1999, 18-year-old Jesse Gelsinger, who had enrolled voluntarily in a clinical trial at the University of Pennsylvania, died suddenly of multiple organ failure after his immune system over-reacted catastrophically to the agent he was given. The product under trial was an engineered adenovirus carrying a gene to correct the serious but rare liver disorder Gelsinger had been born with that leads to the build-up of ammonia in the bloodstream. After his death, the FDA temporarily suspended clinical trials of all gene therapy and subsequently tightened the rules on safety precautions. These were difficult times for pharmaceutical companies developing such agents, and in 2003 Onyx sold the licence for ONYX-015 to the Chinese company Shenzhen Si Biono Gene Technologies Ltd.

In the meantime, China itself had been developing an oncolytic agent very similar to ONYX-015. Oncorine, manufactured by Shanghai Sunway Biotech Ltd, was the first engineered oncolytic virus worldwide to reach the medicine cabinet when it was approved by the Chinese regulatory authorities in 2005. This approval raised the spirits of the depressed gene-therapy field. Today Oncorine is used in China in conjunction with chemotherapy (as an alternative to Gendicine) to treat tumours of the head and neck, and the data from trials suggest it is roughly twice as effective as chemotherapy alone. The goal of the Chinese companies is still to obtain approval for ONYX-015 or Oncorine for use in the US and Europe.

However, the propensity of the viral vector to be detected and wiped out by the patient's immune system before it can deliver its cargo to the cancer cells remains a major challenge for scientists working to refine gene therapy. Another challenge is to find ways of reaching the scattered metastases with these drugs, for it is these secondary tumours that tend to kill the patient with cancer.

SMALL MOLECULES KICK-START STRESS RESPONSE

Other new strategies for treatment being explored start with the fact that in very many cancers p53 is not mutant, but the normal protein is inactivated by some other mechanism. In cervical cancer, for example, around 90 per cent of cases are caused by infection with human papilloma virus (HPV), a sexually transmitted disease that can also cause genital warts. Scientists at the US National Cancer Institute discovered in 1990 that one of the viral genes in cells infected with HPV produces a protein called E6 that completely stymies the action of p53. 'What happens,' explained Karen Vousden, one of the NCI team at the time, 'is that E6 binds to p53 along with some other proteins. The end result is that the p53 protein is degraded very rapidly – it's just broken up into little bits – so the cell never manages to make any p53 protein that's functional. It's as though there isn't any p53 at all.' Preventing infection with HPV in the first place was the obvious solution here, and a vaccine capable of doing just that was approved for the market in autumn 2005.

But the HPV story is an unusual one. More typically the normal p53 protein is prevented from carrying out its functions by abnormalities elsewhere in the tumour-suppression pathway, such as over-zealous behaviour on the part of its controller, Mdm2. This, you will remember from Chapter 13, is the gene switched on by p53 that produces a protein that, in its turn, binds to p53 protein and marks it up for destruction. This dance of death between p53 and Mdm2 goes on in an endless cycle, taking about 20 minutes to complete each time. In this way, Mdm2 ensures that the enormously powerful p53 – with its ability to kill cells or stop them dividing – is kept in check until needed. Once researchers began to understand this feedback mechanism, they figured that if they could release p53 from the clutches of Mdm2 by blocking the interaction between the two proteins, they should be able to reactivate normal p53 in cells where it was abnormally restrained – that is, cancer cells. In such cells, they reasoned, p53 is like a loaded gun primed to go off, but with the trigger jammed; the challenge was to find something that would release the trigger.

Working in his Dundee lab, David Lane became, in the late 1990s, the first person to manage to do this, with a tiny molecule that plugged the docking site between p53 and Mdm2. Because of the awkward shapes of proteins in general and the flexibility of p53 in particular, this was a huge technical challenge, Lane commented when I spoke to him for this book. 'As tough as the search for antiretroviral drugs for HIV?' I queried. 'HIV is a very instructive example, actually,' he replied. 'When I was growing up as a young microbiologist, I was told there would never be a drug to treat a virus; it would be impossible, because the viruses are so close to the host and they use the host machinery. I was also told you'll never get a drug that inhibits a protein–protein interaction. And the immune system will never have a role in helping to clear cancer cells. So you know, you get told these things with absolute certainty by people . . . And of course they're always wrong!' he laughed.

Today there are at least half-a-dozen drugs in development that disrupt the bond between p53 and Mdm2. The earliest and best known is Nutlin, produced by the pharmaceutical giant Hoffmann–La Roche since 2004 (the drug derives its name from the company's research institute in Nutley, New Jersey, where it was developed). But the initial excitement generated by Roche's success soon gave way to serious concern. The first reports of Nutlin described experiments with cells and tissues growing on gels in Petri dishes in the lab. But a mouse experiment reported in 2006 that uncoupled p53 from its controller had a catastrophic outcome that gave everyone working with this strategy pause for thought.

Having engineered a mouse with the gene for the p53 controller Mdm2 knocked out, scientists in Gerard Evan's lab gave the animal a drug to switch on the tumour suppressor. With no controller, the p53 protein went into overdrive in cells throughout the body, resulting in mass, generalised apoptosis – effectively a mouse that melted. In fact, the mouse with no Mdm2 at all was not a relevant model for Nutlin and other such drugs, which are designed to uncouple p53 from its controller only transiently. However, it did raise questions about how to limit the destructive activity of p53 to the tumour sites. This is something drug developers have worked hard on, and modern versions of the drug show very little activity beyond their target cells.

In the lab, Nutlin has inhibited the growth of cells taken from a wide range of cancers, including colon, lung, breast, skin and blood, and it has shown activity in animal models too. But the results have been puzzlingly inconsistent, says Lane, who works closely with the Roche team on the continuing development of Nutlin. In recent experiments with acute myeloid leukaemia (AML), for example, they treated cancer cells from a number of different patients in Petri dishes; they found that while all the cells went into growth arrest at the same low dose of the drug and some committed suicide, it took 10 to 20 times the 'normal' dose to induce apoptosis in others. What was going on? No one is certain yet, though a reasonable hypothesis is that the balance between pro- and anti-suicide proteins active in a cell at the time of treatment affects its sensitivity. Until they understand fully the forces at work, the drug developers will be unable to say exactly which patients with AML should be treated with Nutlin and at what dosages to induce the desired effect, cell death, says Lane.

Meanwhile, researchers are investigating the use of Nutlin in combination with conventional chemo- and radiotherapy, and have found it to be more effective than using either Nutlin or conventional therapy alone in a number of tumour types. One objective of combination therapy is to harness the synergy between the different agents in order to be able to reduce the dosage of the conventional drugs – and thus their distressing side effects for patients – without reducing effectiveness. This is a pressing need in the case of sarcomas, which include bone cancers and are among the most common cancers in children, and here Nutlin looks promising. In 2011, scientists trying to kill sarcoma cells in the lab found they could reduce the amount of some conventional drugs by a factor of 10 when they used them in combination with Nutlin and still achieve the same or a better result as when they used the chemotherapy drugs alone.

Breaking the bond between p53 and its controller Mdm2 is such an attractive option for drug developers that a number of big international pharmaceutical companies, including Merck and Sanofi, and several smaller ones are in the race to get a drug of this kind into the clinic. Until they can do so, however, they still need definitive answers to the vital questions: what effect do drugs of this nature have, if any, in normal cells and exactly how toxic might they be?

Galina Selivanova at the Karolinska Institute in Stockholm is working on a drug of this design which she has named RITA. She points out that in order to kill cells, it is generally not enough for p53 simply to be present; to become active, the tumour suppressor needs to receive clear signals that the cell is under stress – signals that are likely to be strongest in cancer cells. 'My hope is that if you have an Mdm2 inhibitor which is not too strong – maybe it's enough to release just some p53 from Mdm2 – it will not have very drastic effects in normal cells. But in tumour cells, where you have all these signals which are activating p53, it will kill.'

MENDING THE MUTANT

When she left Russia in 1992 with a PhD in bacterial genetics from Moscow University, Selivanova intended to spend just three months of summer gaining experience with work on higher organisms before returning home. However, she joined the lab of Klas Wiman at the Karolinska Institute, discovered p53 and never went back. 'It was so exciting from the start,' she told me when I met her at a mutant p53 meeting in Toronto. 'p53 is *unbelievably* interesting. Everything you do opens new questions, new perspectives.' She joined the p53 community just as people were beginning to think seriously about translation – how they might use the wealth of knowledge they had accumulated to improve the treatment of people with cancer. It was a topic with personal significance: Selivanova had seen her own mother die of a brain tumour, and she soon found herself drawn into the quest.

Besides her own work with RITA, she and Wiman have worked together on another drug, known as PRIMA-1, that is turning out to be one of the most exciting p53-based therapies in development. The drug is designed to work in cancer cells where p53 is mutant and the protein it produces misshapen so that it cannot bind to DNA, as it should, in order to switch on other genes. PRIMA-1 is able to restore the mutant protein to its normal shape, and serendipity played a large part in its discovery. In 1995, the two scientists were studying small scraps of protein called peptides, looking for ones that could regulate the activity of p53. They were intrigued to discover one peptide that was able to activate both normal and mutant p53.

This was clear evidence that 'mending' mutant p53 was possible, and Selivanova was very excited. 'It was fantastic,' she recalled with a smile. 'I wanted, of course, to go out and cure tumours – at least in mice!' But the peptide proved unworkable as a drug: in a living organism these scraps of protein are poor at entering cells and are quickly broken down and recycled. What was needed was a chemical compound, a small molecule that would perform the same tricks.

Working with Wiman and a new postdoc, Vladimir Bykov, Selivanova screened thousands of compounds from a library of possible candidates provided by the US National Cancer Institute. In 1999, the three discovered a molecule they named PRIMA, an acronym for 'p53 reactivation and induction of massive apoptosis' that appealed to the scientists because it also implies something that is first class. Experimenting with the molecule they found, to their gratification, that it is effective with a wide range of p53 mutations, and therefore potentially useful in treating many different tumour types. They published their results soon afterwards, 'and PRIMA attracted *a lot* of media attention,' Wiman, a tall, soft-spoken Swede, told me when I visited him at the Karolinska Institute. 'I was on TV and in newspapers and journals around the world, because the concept of having a small molecule that will make the cancer cells commit suicide is so appealing.'

So how does PRIMA-1 work? Wiman and Bykov discovered, to their surprise, that both PRIMA-1 and a very similar compound known as PRIMA-1 MET are converted to another compound that binds tightly to p53 protein and refolds it. 'This was a very important and exciting finding since it gave us a better understanding of how these compounds can reactivate mutant p53,' said Wiman.

In partnership with the Karolinska, he, Bykov and Selivanova set up a small biotech company to develop PRIMA-1 MET for the market. It has been a steep learning curve. 'As scientists you need to work with company people – a completely different culture,' commented Wiman. 'Suddenly there are people in suits, board meetings, talk about money . . . And then you interact with clinicians too. So there are three worlds and you all have to work together all the way through. We had no idea what was involved when we started.'

The company has taken PRIMA-1 MET through a phase 1 clinical trial, which involved 22 patients with cancer of the prostate and blood being given a short course of the drug by injection. Phase 1 trials are designed to test patients' tolerance to a potential new drug, and to find out how it disperses in the body and how long it persists. The results, published in 2012, were promising: they showed that PRIMA-1 MET is not toxic and that side effects – including dizziness and fatigue – are mild.

Phase 2 clinical trials, designed to prove that the drug works in people as it does in the lab and in animal models, are the next step. The Karolinska researchers and their company are hoping to test PRIMA-1 MET in combination with conventional chemotherapy in cancer patients, where the two drugs are expected to act in synergy: while PRIMA-1 MET restores mutant p53 to its normal shape and function, the other drug will cause DNA damage that sends clear signals of stress to trigger apoptosis. But this is where the hurdles en route to the clinic really begin: staging a phase 2 trial for PRIMA-1 MET is likely to cost millions of euros, said Wiman. A tiny biotech company like theirs needs to find a partner with serious money to invest.

A doctor in Denmark who has seen the effects of PRIMA-1 MET on lung-cancer cells in the lab and in mice is so excited by the drug that he has offered to set up a trial himself. The MD Anderson Center, too, is keen to run a phase 2 trial of PRIMA-1 with cancer patients. But finding the funds for all these activities remains a huge challenge, and so far Big Pharma has shown little interest in small molecules that restore normal shape and function to mutant p53 because it is still not entirely clear how they work.

SMART THERAPY

A problem that dogs the field of cancer therapy is the issue of drug resistance. The extreme instability of cancer cells and the terrible speed with which they pick up mutations mean that they are likely to find a way round a targeted drug before too long, no matter how clever the design, as the cells that survive the initial onslaught of treatment give rise to equally hardy clones that grow into resistant tumours. To minimise the prospect of failure, oncologists typically treat their patients with a combination of therapies – either a cocktail of drugs, or a drug together with radiotherapy. With this strategy, cancer cells that are not affected by one drug should be hit by the other.

Researchers are also investigating the use of drug combinations in a novel kind of p53-based treatment called cyclotherapy. One of the biggest shortcomings of conventional chemotherapy, which is 'cytotoxic' (meaning that it's a cell poison) and targets the body's rapidly dividing cells, is that it is indiscriminate. Cancer cells are by definition fast-dividing, but so too are the cells in the hair follicles, lining of the gut and bone marrow, which sustain collateral damage during chemotherapy. But hair loss, nausea, diarrhoea, anaemia and depletion of the immune system are not just distressing side effects for the patient, they are potentially deadly and they limit the dose of cytotoxic drugs the oncologist can administer to attack the cancer.

The principle behind cyclotherapy is that patients be given one drug to 'protect' the healthy cells from the chemotherapy by temporarily stopping them from dividing, while their cancer cells (which continue to divide and therefore remain targets of the chemotherapy) are blasted with a second, cytoxic drug given simultaneously. With healthy cells protected, the theory goes, the oncologist will be able to increase the dose of the cytotoxic drug and thus maximise its potential to wipe out the tumour. But even if it falls short of clearing the cancer completely, cyclotherapy will make chemo a lot less unpleasant for the patient because it will limit the side effects by sparing the cells of the hair, gut, bone marrow, etc. from the full force of treatment.

In laboratory tests, Nutlin is looking the most promising of a number of similar drugs used to protect the healthy cells. However, cyclotherapy is still a few years from the clinic. Researchers still need to work out which combinations of drugs work best, with what tumour types and in what quantities. The arrest of healthy cells mid-cycle must be reversible: too high a dose of the protective drug, for example, could cause healthy cells to senesce, but too low a dose might not arrest them for long enough to protect them from the cytotoxic drug. And no one is sure yet how well cyclotherapy works in living organisms: as of 2012 there was only one published report of an experiment in mice. However, one of the main constraints on cyclotherapy is the fact that neither Nutlin nor any of the other potential 'protectors' has yet been approved for use in the clinic in its own right.

NEW LIGHT ON OLD TREATMENT

Despite the frustratingly slow progress of brand new p53-based therapies, scientists' understanding of p53 is already beginning to have an impact on the treatment of cancer patients: it enables oncologists to make more rational decisions about the use of conventional chemo- and radiotherapy.

Chemotherapy has a colourful, if unfortunate, history. Its origins go back to World War I, when the Germans used mustard gas in the trenches of Europe to devastating effect. The use of chemical weapons was banned by the Geneva Protocol of 1925, but not the possession of such weapons, and the Americans continued to develop and stockpile them. In December 1943, a US cargo ship, the SS *Harvey*, secretly carrying mustard-gas bombs to the Mediterranean war front, was sunk in a German raid on the port of Bari, southern Italy, and a cloud of gas drifted over the city. No one knows how many civilians were affected, but more than 600 military personnel were hospitalised and 83 died. During autopsies of the victims, pathologists found evidence that the normally fast-dividing cells of the bone marrow and lymphoid tissues had been suppressed. From this observation came the idea that perhaps such an agent could be used to attack the rapidly dividing cells of cancer.

Soon scientists were doing experiments with mustard gas in mice. Encouraged by the results, they moved cautiously on to testing the agent in humans. The first human subject was a patient with lymphoma – cancer of the lymphoid tissue – and his doctors observed with delight the dramatic shrinkage of his tumours after administration of the drug. Unfortunately the effect was short-lived, but it galvanised the cancer community: here at last was a new way of treating the disease. For many centuries, surgery had been the only option for getting rid of tumours, and patients' long-term survival chances were minimal.

Over the decades since, many different cytotoxic drugs have been developed – all on the same principle, that they are poisonous to cells. But while chemotherapy has been found to work wonderfully well in some tumours, it does not work at all in others. And in some it works for a while and then stops. Why is the response so varied? Until p53 research began offering clues, no one had an answer. Today we know that both chemo- and radiotherapy work not by killing cancer cells directly, in a sledgehammer kind of way as had been assumed, but much more subtly: typically, these therapies work by inducing cancer cells to commit suicide in response to damage of their DNA – the normal response to cell stress, mediated by p53.

Scott Lowe, whom we met in Chapter 12 creating mouse models and making groundbreaking discoveries about apoptosis and p53, was one of the first to recognise the tumour suppressor's central role in conventional therapy. To recap, Lowe subjected the highly sensitive thymus glands of his mice to radiation and discovered that the cells with normal, functioning p53 died very quickly by apoptosis, but cells with mutant or no p53 were resistant to radiation and survived. Confirming p53's role in apoptosis in response to radiation set Lowe wondering more generally: could p53 be responsible for the effect of radiation – and perhaps cytotoxic drugs also – in cancer therapy? As an idea, it was incredibly simple, and so obvious in retrospect, but revolutionary at the time.

'Here was a situation where the hypothesis was that if p53 is mutant, the standard chemotherapy drugs are less likely to work,' explained Lowe. 'In the case of leukaemias and lymphomas what we would have predicted holds true. But now, 17 years of subsequent research says of course it's more complicated than that.'

In leukaemia and lymphoma cells, p53 is almost always normal and, as one would expect, these cancers are highly sensitive to chemo- and radiotherapy. But in solid tumours (cancers of the organs rather than the blood), the picture is much less predictable – and sometimes it is counter-intuitive. In some types of cancer, cells with mutant p53 are more responsive to cytotoxic drugs than are cells with normal p53. This is the case with glioblastoma, an aggressive tumour of the brain, for example. So what is going on?

One explanation is that in these cases, the cells with mutant p53 are indeed killed in the sledgehammer way oncologists originally imagined. They sustain severe damage to their DNA that cannot be repaired because p53 is out of action, nor can cell division be arrested. The cells carry on chaotically through the cycle and eventually succumb to what is called 'mitotic catastrophe' – wholesale failure of the machinery of replication. This scenario implies that it is essential for oncologists to know which way a tumour type will react to conventional therapy, depending on its p53 status. But things can get even more complicated.

In some cases, giving chemo- or radiotherapy to patients whose cancers have *normal* p53 can actually make things worse. Cell death, as we know, is just one of several options chosen by p53 in response to damaged DNA. It can also choose to arrest the cell mid-cycle and send in the repair team before releasing the cell to carry on replication. Or it can condemn the cell to senescence – permanent arrest, which we know from the chapter on ageing can eventually stimulate cancer in neighbouring cells. Thus cancer cells that are not killed by chemo- or radiotherapy can be the seed stock for further tumours – and sometimes these new tumours are especially aggressive simply because the cells are survivors of highly toxic treatment, and bred for resistance.

This makes sense, but it is only a hypothesis at present – there are no experimental data to prove it definitively. One source of confusion is the fact that, in the vast wealth of research that is carried out on p53, there is so little consistency in the methodology that it is hard to compare results. 'In experimental systems we have all kinds of effects,' said Pierre Hainaut. 'You can always get an experimental system to behave as you would like it to, as an investigator! Now if you go to real life . . .' Hainaut sat me down in front of the computer in his study at his Lyon home and brought up a paper he was about to submit. It was an analysis of a number of clinical trials involving the use of a common chemotherapy drug, Cisplatin, in lung-cancer cases. Overall, the effect of the drug was small, but he and his colleagues wanted to know whether the p53 status of an individual patient's tumour influenced the outcome of Cisplatin treatment. For their analysis they had before them the biggest data set of its kind. It contained information about the outcome of treatment, plus the p53 status of the tumours, for 1,200 cancer patients from four trials, conducted in Canada, the US and Europe.

The researchers found – unsurprisingly – that patients whose tumours had normal p53 did a lot better than those with mutant p53. But what did surprise them was that patients with some specific mutants – but, crucially, not others – got dramatically worse after Cisplatin treatment. Their tumours spread aggressively and many patients died even more quickly than they would likely have done with no treatment at all. Hainaut was not certain, at that point, whether it was the metastases that killed the people – he was awaiting further information from a statistician – but that was his hunch.

Whatever the final cause of death turns out to be, knowing the p53 status of lung tumours will be useful in deciding who should receive Cisplatin therapy and who should not. 'We are not doing well with lung cancer,' Hainaut reflected as he scrolled through his paper on the computer screen. 'There are probably 1.5 million people in the world with this type of cancer. Maybe 500,000 receive this treatment every year – and they receive it "blind", because p53 is not being tested by mutation in these patients up front. Such a test would clearly improve the outcome. It would be really worthwhile . . . That's the lesson of our analysis.'

The situation Hainaut was describing was specific to lung cancer, with certain p53 mutations, treated with Cisplatin. But the lesson holds true more generally. What scientists have discovered about p53 and its role in conventional therapy offers cancer specialists a tool for making more rational decisions about how best to treat their patients. This is especially true when p53 status is part of a wider analysis of the genetic make-up of a tumour, because so many things besides this tumour suppressor have an impact on treatment. At present such tests are rarely offered in cancer clinics, but things are changing fast. As full genome sequencing becomes ever easier, quicker and cheaper to perform – and as the new gene therapies that target the defects specific to an individual patient's tumour begin to reach the clinic – genetic analysis will become a routine part of diagnosis and treatment. Genetic analysis is an essential part, too, of the latest strategies for cancer prevention.

THE BEST CHANCE OF SURVIVAL

Compared to new treatment ideas, prevention studies have a tough time attracting cancer-research funds. The science of prevention is not as sexy; it doesn't offer the same rewards to Big Pharma; and besides, it's easier to get excited about tumours that are cured than about tumours that just don't happen.

Nevertheless, Bert Vogelstein is not deterred. 'We believe the major impact on cancer over the next half-century will come not from treating *advanced* cancers, but from preventing cancer – in particular from detecting tumours at a very early stage,' he said. 'Virtually all cancers are treatable by surgery, without the need for any chemotherapy or radiation, if they're caught early enough. That's definitely true for colon, but it's also true for many other tumours. It's an underlying principle.'

For a number of years now, Vogelstein's lab at Johns Hopkins has been busily engaged in developing tools to look for evidence of early cancers. They are focusing their efforts on detecting biomarkers – bits of mutant DNA sloughed off by cancer cells that might be swilling around in a sea of normal DNA molecules in the blood, urine, stools or sputum, bearing witness to the presence of furtive disease. 'The best marker, the best gene, is obviously p53, because it's mutant in more tumours than any other gene – that's the bedrock of this test,' explained Vogelstein.

The body fluid in which a biomarker is found is often a good indicator of where the tumour is developing: urine suggests bladder cancer, for example, stool suggests colon cancer and sputum suggests lung. By late 2012, Vogelstein's team had investigated more than 700 cancers, starting with advanced tumours, to see if they could find free-floating biomarkers. 'In advanced cancers of most tumour types – that is breast, colon, pancreas, lung – you can detect well over 90 per cent of them in the blood,' he commented. For advanced colon cancer the researchers' detection rate in stool samples is close to 100 per cent, and even in relatively early, pre-metastatic cancers it is 85–90 per cent. 'This test is starting to rival colonoscopy in sensitivity,' said Vogelstein. He reckons that even in blood samples, his team has more than a 50/50 chance of detecting colon cancer before it has spread. 'And if you can detect even 50 per cent of cancers at a stage when they're curable that would be massive.'

Researchers working on the problem of liver cancer in West Africa have found biomarkers in the blood that can be used to screen for the disease before symptoms arise. In this region, you will recall, liver cancer is often associated with aflatoxin contamination of food crops, and DNA molecules released into the blood from a diseased liver show the characteristic fingerprint mutations in the p53 gene. Elsewhere, too, scientists are exploring the possibilities of using the presence of mutant p53 in body fluids to screen for early cancers.

***

Such screens have still to be widely validated and refined before they reach the clinic. However, many scientists working on the front line of p53 believe we are on the threshold of a golden age in cancer prevention and cure. In coming years we should expect to see:

- gene therapy become routine treatment for cancer as researchers perfect the technique for modifying viruses as delivery vehicles (dozens of people with diverse genetic disorders have already been treated successfully);
- more widespread use of genetic analysis of tumours, and the status of p53 being used to determine the best course of treatment and to predict outcomes and long-term prognosis;
- a dramatic decrease in side effects of cancer therapy as treatment becomes more accurately and exclusively

targeted at tumour cells, and as strategies such as cyclotherapy are used to protect the body's normal cells;
- a variety of p53-based drugs for use in different circumstances that are able to manipulate the tumour-suppressor pathway to kill cancer cells.

'I'm very, very optimistic,' said Gerard Evan. 'I think we're going to see dramatic shifts in our ability to treat and contain human cancers over the next 10, 15, 20 years.' And he added, perhaps provocatively, 'My daughter is 22 and my son is 21, and I can pretty confidently say they will never, ever have to worry about dying from cancer.'

Dramatis Personae

Note: these thumbnail biographies do not include all the people mentioned in this book, but are intended as an aide-memoire to some key players whose names appear frequently and sometimes out of context.

EVAN, GERARD
A scientist with Cancer Research UK (CRUK), now based at Cambridge University as Professor of Biochemistry. An early enthusiast for the use of mouse models to find out how things work in living organisms, he is renowned as an original thinker whose work frequently challenges mainstream thinking. We meet him first in Chapter 1, making the provocative comments about the rarity of cancer.

HAINAUT, PIERRE
Based at the World Health Organization's International Agency for Research on Cancer (IARC) in Lyon, France, for many years, Hainaut ran the mutant p53 database – a detailed record of all the different mutants appearing in the literature and how they behave. A natural-born detective, he has a special interest in tracing the distribution and pattern of disease caused by mutant p53 among people throughout the world.

HALL, PETER
A Professor of Pathology and close colleague of one of p53's discoverers, David Lane, at Dundee University, Scotland, in the 1990s. p53 has been one of Hall's main research interests. Especially renowned for the maverick experiment he cooked up with Lane in a Dundee pub to test the effects of radiation on p53 in living organisms.

KNUDSON, ALFRED

The US-based cancer geneticist who first hypothesised the presence in our cells of genes whose job is to protect us from cancer. The 'two-hit' hypothesis of 1971, which grew out of Knudson's work with children with the eye tumour retinoblastoma, changed forever the way in which cancer biologists viewed the process of tumour formation.

LANE, DAVID

A central character in the story of p53 as one of four people who, working entirely independently, discovered the gene in 1979. Lane was then at the ICRF (Imperial Cancer Research Fund, now known as Cancer Research UK) in London. At Dundee University in the 1990s, he built one of the largest communities of scientists working on p53 anywhere in the world. Responsible, among many other things, for dubbing p53 'Guardian of the Genome'.

LEVINE, ARNIE

The Princeton-based scientist who discovered p53 independently but at the same time as David Lane and two others in 1979. Levine's lab has been a hub of p53 research ever since and has been involved in many of the most important discoveries about the function of the gene.

OREN, MOSHE

One of the first people, in 1984, to make a clone – an exact replica – of p53 from which endless copies could be made for research. Among many other important contributions to the field, Oren also discovered p53's role in apoptosis (cell suicide) and, along with Arnie Levine and Carol Prives, helped to uncover the mechanism that keeps the powerful p53 under strict control in our cells.

PRIVES, CAROL

A Columbia University-based scientist who, in collaboration with Bert Vogelstein, discovered that p53 works as a master switch in our cells, turning other genes on and off in response to signals. Involved also, along with Moshe Oren and Arnie Levine, in discovering the mechanism that keeps the powerful p53 itself under strict control in our cells. Prives is among a number of key researchers at the core of the p53 community.

ROTTER, VARDA
Based at the Weizmann Institute in Israel, Rotter was one of the earliest researchers to recognise that p53, when mutated, does not simply lose the ability to function as a tumour suppressor; very often the mutant acts to promote the growth of a tumour. Rotter is famed for having stuck to her guns, even when her analysis was challenged by some of the best brains in the community, and today hers is the mainstream view.

VOGELSTEIN, BERT
Trained as a medical doctor at Johns Hopkins in Baltimore, Vogelstein's experience treating children with cancer led him to molecular-biology research. The first person to investigate p53 in human cancers, he has been involved in many key discoveries about the function of the gene, including its role as a tumour suppressor and a master switch.

WEINBERG, ROBERT
Eminent US scientist involved since the early days of the molecular-biology revolution in uncovering the genetic basis of cancer. Best known for his discoveries of the first human oncogene (or cancer-promoting gene) and the first tumour suppressor. Weinberg has spent most of his working life at the Massachusetts Institute of Technology (MIT) and is the author, with Doug Hanahan, of a seminal paper, 'The Hallmarks of Cancer', which defines the key characteristics of all cancer cells.

WYLLIE, ANDREW

Trained as a pathologist, Wyllie was a PhD student at Aberdeen University in Scotland when 'programmed cell death', or cell suicide, emerged from rarefied fields into mainstream biology and was given the name 'apoptosis'. His was one of two groups of researchers who simultaneously discovered that apoptosis is one of the programmes p53 is able to trigger in response to cellular stress in real life, not just in cell cultures in Petri dishes.

Glossary

Allele: One of a pair of genes that occupy the same site on a chromosome. All genes come in pairs: you inherit one allele of each gene from your mother and one from your father.

Antibody: Antibodies are the soldiers of the immune system; they move freely in the blood, seeking out invaders such as bacteria and viruses and flagging them up for destruction. Antibodies are tailor-made by the immune system to recognise and attach to specific targets, which makes them excellent tools for 'finding' target molecules in researchers' laboratory experiments.

Apoptosis: Programmed cell death, or cell suicide.

Bacteriophage: A virus which targets and infects bacteria.

Carcinogen: A substance capable of causing cancer.

Carcinoma: A type of cancer that starts in the epithelial cells that form the outer membranes of all the organs, tubes and cavities in our bodies, and include our skin. At least 80 per cent of cancers are carcinomas (see also **sarcoma, leukaemia, lymphoma**).

Cell culture: A laboratory process in which cells are maintained and grown outside the body in specially designed containers, such as test tubes and Petri dishes, and under precisely controlled conditions of temperature, humidity, nutrition and freedom from contamination.

Cell line: A cell culture developed from a single cell and therefore consisting of cells with a uniform genetic make-up.

Cell-cycle checkpoint: The checkpoints mark the end of each phase in the multi-phase process of cell division. At each checkpoint, 'quality control' has the chance to verify that the process has been accurately completed before allowing the cell to proceed to the next phase.

Checkpoint: See **cell-cycle checkpoint**.

Codon: A sequence of three consecutive nucleotides (the basic building blocks of **DNA**) on a gene that together form a unit. These units dictate which amino acids are to be used to create the protein that will carry out the function of the gene.

Clone: In the context of this book, a gene that is produced artificially from another gene, of which it is an identical copy.

DNA: Deoxyribonucleic acid, the material inside the nucleus of the cells of living organisms that carries genetic information.

Expression: The process by which an activated gene makes a protein or other product that carries out the function of that gene in the cell. If a gene is 'over-expressed', it implies there is an over-abundance of protein in the cell.

Gain of function: An expression used in reference to a genetic mutation that changes the gene product (e.g. protein) in such a way that it gains a new and abnormal function (see also **loss of function**).

'Hallmarks of Cancer': A seminal paper written by Robert Weinberg and Doug Hanahan in 2000 that describes the six characteristics common to all cancers, of whatever organ or origin. They revised the 'Hallmarks' in 2011, adding four more general principles.

Large T antigen: The gene in the **DNA** of the monkey virus SV40 that is responsible for causing cancer in the cells of the host species it infects.

Leukaemia: Cancer of the white blood cells, which are a vital component of the immune system (see also **lymphoma**).

Loss of function: An expression used in reference to a mutation that renders a gene useless – the mutant gene is either unable to make any protein or the protein it makes has no function. In most, if not all, tumour-suppressor genes other than p53, mutation leads to 'loss of function'.

Lymphoma: Cancer originating in lymphoid tissue, a key

component of the body's immune system. Cancers of lymphocytes (lymphomas) and other white cells in the blood (**leukaemia**) together account for about 6.5 per cent of all cancers.

Malignant: In medical usage malignant means cancerous; able to spread to other parts of the body.

Metastasis: The spread of cancer cells from the original site to other parts of the body (hence **metastases**: secondary cancers).

Mutagen: A substance capable of causing mutation.

Mutant: Something that has undergone mutation (see below).

Mutation: A change of the **DNA** sequence within a gene or chromosome of an organism resulting in a new character or trait not found in the parental type; or the process by which such a change occurs.

Nucleotide: Nucleotides are the basic building blocks of **DNA**, which stack one on top of the other like nano-sized blocks of Lego to form the long ribbons of the double helix.

Oncogene: A gene that has the potential to cause cancer. Very often these are genes that have a normal role to play in the growth of cells, but that have sustained a **mutation** and lost the ability to respond to control signals.

Oncogenic: Causing development of a tumour or tumours.

Oncology: The study of cancer (hence **oncologist**, a doctor or scientist specialising in cancer).

'Postdoc': A postdoctoral scholar; an individual with a doctoral degree who is engaged in a temporary period of mentored research and/or scholarly training in order to acquire the professional skills needed for his or her future career.

Recombinant DNA: DNA that has been formed artificially by combining genetic material from different organisms.

Sarcoma: A type of cancer that forms in the connective or supportive tissues of the body such as muscle, bone and fatty tissue. Sarcomas account for less than 1 per cent of cancers.

Senescence: In this book the term is used to describe a state in which cell is no longer able to divide but remains alive and functioning.

Somatic mutation: A **mutation** in a mature cell that has occurred spontaneously during the course of life, as opposed to one that is inherited and will be present in all the cells, both normal and cancerous.

Tissue culture: The growth of tissues or cells removed from an organism. The living material is placed in a lab dish such as a test tube or Petri dish with a growth medium, typically a broth or agar gel, that contains special nutrients.

Transcription factor: A protein that binds to **DNA** at a specific site and controls the expression of a gene or genes in the vicinity, switching them on and off as appropriate.

Transformation: In this book, this term is used to describe the process by which a cell acquires the properties of cancer (commonly described also as 'malignant transformation').

Tumour suppressor: A gene whose function is to prevent cells from becoming malignant.

Wild type: Used in reference to a gene, this means the 'normal' gene that functions as nature intended, as opposed to the '**mutant**' gene whose behaviour is aberrant.

Notes on Sources

Besides my personal interviews with many of the key players in the p53 story, I have tapped into a rich repository of information contained in a great number of books, journals and multimedia websites in my research for this book. I often used multiple sources for a single discussion point, and list here those I found particularly useful and that deserve special mention in each chapter. Some sources, however, have provided information, insights and ideas of relevance throughout the book. They include:

Hainaut, Pierre, & Wiman, Klas (eds), *25 Years of p53 Research* (Dordrecht, Netherlands: Springer, 2005)

Judson, Horace Freeland, *The Eighth Day of Creation: Makers of the Revolution in Biology,* (Woodbury, NY: Cold Spring Harbor Laboratory Press, 1996)

Lane, David, & Levine, Arnold, *p53 Research: The Past Thirty Years and the Next Thirty Years* (Woodbury, NY: Cold Spring Harbor 'Perspectives in Biology', May 2010). This is one among many excellent papers I drew upon from Cold Spring Harbor's 'Perspectives in Biology' collection, available at *cshperspectives.cshlp.org/cgi/collection/* (*see:* The p53 Family).

Mukherjee, Siddhartha, *The Emperor of All Maladies: A Biography of Cancer* (London: Fourth Estate, 2011)

Varmus, Harold, *The Art and Politics of Science* (New York: W W Norton & Co., 2009)

A Conversation with Robert Weinberg (from the 'Conversations with Scientists' series sponsored by the MIT Department of Biology and the Howard Hughes Medical Institute). Available at *video.mit.edu/watch/a-conversation-with-robert-weinberg-4508/*

Milestones in Cancer, a series of authoritative articles provided by the science journal *Nature*, available at *www.nature.com/milestones/milecancer/index.html*

The p53 Website. A resource for scientists working on p53 and cancer research set up by Thierry Soussi in 1994, available at: *http://p53.free.fr/*

Preface

The epigraph from Gerard Evan comes from my interview with him in Cambridge, England, in June 2012.

Chapter 1: Flesh of our Own Flesh

The epigraph from Peyton Rous comes from his lecture to the Nobel Committee on winning the prize, 'The Challenge to Man of the Neoplastic Cell', available at *www.nobelprize.org/nobel_prizes/medicine/laureates/1966/rous-lecture.html*

Besides the *Conversation* recorded at MIT and cited above, Robert Weinberg spoke of his work with Doug Hanahan on the Hallmarks of Cancer at a conference of the National Cancer Research Institute in 2010. Available at *www.youtube.com/watch?v=RP4js-yYK2U)*

Chapter 2: The Enemy Within

The epigraph from Michael Bishop comes from his book, *How to Win the Nobel Prize: An Unexpected Life in Science* (Harvard University Press, 2003), page 161.

For information on Peyton Rous, I relied on the excellent archives of the Nobel Foundation, see: *http://www.nobelprize.org/nobel_prizes/medicine/laureates/1966/rous-bio.html*

Besides their autobiographical books already cited, the Nobel archive also was a rich source of information on Varmus and Bishop, who won the prize in 1989. See *www.nobelprize.org/nobel_prizes/medicine/laureates/1989*

For the Asilomar debate see M. J. Peterson, 2010, *Asilomar Conference on Laboratory Precautions*. International Dimensions of Ethics Education in Science and Technology. Available at *www.umass.edu/sts/ethics*

Chapter 3: Discovery
The epigraph comes from Judson's book, *The Eighth Day of Creation,* cited above, page 10. The footnote quote is from Jeffrey Taubenberger; see *www.pathsoc.org/conversations*

Chapter 4: Unseeable Biology
The epigraph comes from *A Short History of Nearly Everything* by Bill Bryson (London: Transworld Publishers, 2003), page 451.

For this chapter I relied substantially on the information provided by the National Human Genome Research Institute of the National Institutes of Health, available at *www.genome.gov.*

Chapter 6: A Case of Mistaken Identity
The epigraph from Judson comes from *The Eighth Day of Creation* cited above, page 594.

Besides the conversation recorded at MIT cited above, see Robert Weinberg's description of his discovery of oncogenes in human tumours, available at *http://www.bioinfo.org.cn/book/Great%20Experments/great25.htm*

See also *Nature Milestone 17* and *18* on the discovery of the first human oncogene and on oncogene co-operation *www.nature.com/milestones/milecancer/timeline.html*

Chapter 7: A New Angle on Cancer
The epigraph from the 19th-century French novelist Jules Verne comes from *A Journey to the Centre of the Earth.*

For information on Henry Harris, I relied on the rich archive of the Genetics and Medicine Historical Network set up by Cardiff University with support from the Wellcome Trust. See *http://www.genmedhist.info/interviews/.* See also 'How Tumour Suppressor Genes were Discovered' by Henry Harris in *Journal of the Federation of American Societies for Experimental Biology (FASEB),* Volume 7, pages 978–79.

The most important sources for the finding of the first

tumour-suppressor gene are the MIT conversation with Weinberg cited above and *Natural Obsessions: striving to unlock the deepest secrets of the cancer cell*, by Natalie Angier (Boston: Houghton Mifflin, 1999).

Chapter 8: p53 Reveals its True Colours
The epigraph from Francis Crick, co-discoverer with James Watson of the double-helix structure of DNA, comes from *The Eighth Day of Creation*, page 93.

Suzy Baker tells her story in S. J. Baker (2003), Redefining p53: Entering the Tumor Suppressor Era. *Cell Cycle*, Volume 2, pages 7–8.

Chapter 9: Master Switch
The epigraph from Matt Ridley comes from his book, *Genome: the autobiography of a species in 23 chapters*, page 271. Reprinted by permission of HarperCollins Publishers Ltd © 1999 Matt Ridley.

Besides the personal interviews, this chapter relied heavily on the book *25 Years of p53 Research,* cited at the beginning of these notes.

Chapter 10: 'Guardian of the Genome'
The epigraph from David Lane comes from his commentary, 'Worrying about p53', in *Current Biology*, Vol. 2 (1992), pages 581–583.

Chapter 11: Of Autumn Leaves and Cell Death
The epigraph comes from *The Man Without Qualities,* a novel by the Austrian Robert Musil, unfinished at the time of his death in 1942 and published posthumously (London: Pan Macmillan, 1997, translated by Sophie Wilkins and Burton Pike).

John Kerr's own paper, 'History of the events leading to the formulation of the apoptosis concept', published in *Toxicology*, Volumes 181–182 (2002), pages 471–474, was a key resource.

The interview with Richard Lockshin appears in R. A. Lockshin (2008), Early work on apoptosis, an interview with Richard Lockshin. *Cell Death and Differentiation*, Volume 15, pages 1091–95.

Chapter 12: Of Mice and Men

The epigraph comes from *The Eighth Day of Creation,* page 73.

A key resource for this chapter was the archive of the Nobel Foundation, which awarded the 2007 Prize for Medicine to Mario Capecchi, Oliver Smithies and Martin Evans for their work on transgenic mice. See *www.nobelprize.org/nobel_prizes/medicine/laureates/2007/*

Another important source of information was the National Human Genome Research Institute. See *www.genome.gov/10005834*

Chapter 13: The Guardian's Gatekeeper

The epigraph from Gerard Evan comes from my interview with him in Cambridge, England, in June 2012.

Besides the personal interviews, this chapter relied heavily on the book *25 Years of p53 Research,* cited at the beginning of these notes. See especially Chapter 4: Gatekeepers of the Guardian: p53 regulation by post-translational modification, MDM2 and MDMX, by Geoffrey Wahl, Jayne Stommel, Kurt Krummel and Mark Wade.

Chapter 14: The Smoking Gun

The epigraph from Siddhartha Mukherjee comes from his book *The Emperor of All Maladies: a biography of cancer,* page 241. Reprinted by permission of HarperCollins Publishers © 2011 Siddhartha Mukherjee.

For information on Richard Doll and his research, I drew on two main sources: *Life of A Revolutionary,* by Jonathan Wood, a review of Conrad Keating's biography of Doll that appeared as an 'Oxford Science Blog' from the University of Oxford on 11th November 2009, available at *www.ox.ac.uk/media/science_blog/091111.html*, and Doll's paper in the *British Medical Journal* of 30th September 1950: 'Smoking and carcinoma of the lung: preliminary report' by Richard Doll and A. Bradford Hill.

Information on Angel Roffo is drawn from 'Angel H Roffo: the forgotten father of experimental tobacco carcinogenesis' by Robert Proctor (2006), in *Bulletin of the World Health Organization*, Volume 84, pages 494–96.

A rich source of information for this chapter was the wealth of original documents that the tobacco industry was obliged by US law to make public and which are available online at *tobaccodocuments.org*. See especially *tobaccodocuments.org/profiles/roffo_ah.html* and *tobaccodocuments.org/atc/60359252.html#images*

The Legacy Tobacco Documents Library, developed and managed by the University of California, San Francisco, also has more than 14 million documents available for scrutiny at *http://legacy.library.ucsf.edu/*

A key source for this chapter was the paper by Asaf Bitton and colleagues (including Stanton Glantz) in *The Lancet* of January 2005 (pages 531–540): 'The p53 tumour suppressor gene and the tobacco industry: research, debate and conflict of interest'.

See also the interview with Glantz in *Frontline: Inside the Tobacco Deal*. Available at *www.pbs.org/wgbh/pages/frontline/shows/settlement/interviews/glantz.html*. Copyright © 1995–2014 WGBH Education Foundation. See also *The Cigarette Papers,* edited by Stanton A. Glantz *et al.*, 1998 (University of California Press)

For the p53 database see *http://p53.iarc.fr/*

Chapter 15: Following the Fingerprints

The quotation by Isaac Asimov, American biochemist and science writer, used for the epigraph is of unknown origin, but widely used.

For this chapter I relied on two key documents: 'A role for sunlight in skin cancer: UV-induced p53 mutations in squamous cell carcinoma' by Douglas E. Brash *et al.*, in *PNAS*, Vol. 88, pages 10124–28, November 1991 and 'Sunlight and Skin Cancer' by David J. Leffell and Douglas E. Brash, in *Scientific American*, Vol. 275, 52–59, reproduced with permission copyright © 1996 Scientific American, inc. All Rights Reserved.

Chapter 16: Cancer in the Family
The epigraph from Patricia Prolla comes from my interview with her, Porto Alegre, Brazil, 2012.

Chapter 17: The Tropeiro *Connection?*
The epigraph is from the Polish-born physicist Marie Curie, famed for her pioneering work on radioactivity in the early 20th century.

Chapter 18: Jekyll and Hyde
The epigraph from James Watson, American molecular biologist best known as co-discoverer with Francis Crick of the double-helix structure of DNA, comes from *The Eighth Day of Creation*, page 27.

An important reference document for this chapter was 'Mutant p53 Gain- of-Function in Cancer' by Moshe Oren and Varda Rotter, in *Cold Spring Harbor Perspectives in Biology*, 2010, Volume 2, a001107. Available at cshperspectives.com/content/2/2/a001107.full

For the information on Stanley Prusiner, I relied on the archive of the Nobel Foundation, which awarded him the Prize for Medicine in 1997. See *www.nobelprize.org/nobel_prizes/medicine/laureates/1997*

Other key references were two papers, published simultaneously in the same journal, *Cell*, Volume 119, 2004, by Tyler Jacks and Gigi Lozano and their colleagues: 'Mutant p53 Gain of Function in Two Mouse Models of Li-Fraumeni

Syndrome' by Kenneth P Olive *et al.* (847–860) and 'Gain of Function of a p53 Hot Spot Mutation in a Mouse Model of Li-Fraumeni Syndrome' by Gene A Lang *et al.* (861–872).

See also 'Mutant p53: one name, many proteins' by William A. Freed-Pastor and Carol Prives, in *Genes and Development*, 2012, Volume 26, 1268–86.

Chapter 19: Cancer and Ageing – a Balancing Act?

The epigraph from John Maddox, editor emeritus of *Nature*, comes from his introduction to *The Eighth Day of Creation* by Horace Freeland Judson, page xii.

Key sources for this chapter were 'Using Mice to Examine p53 Function in Cancer, Aging, and Longevity' by Lawrence A. Donehower, *Cold Spring Harbor Perspectives in Biology*, 2009, 1:a001081, see *http://cshperspectives.cshlp.org*; 'Two faces of p53: aging and tumour suppression' by Francis Rodier *et al.* in *Nucleic Acids Research*, 2007, Vol. 35, pages 7475–84; 'Modulation of mammalian life span by the short isoform of p53' by Bernhard Maier *et al. Genes and Development*, 2004, Volume 18, pages 306–19. Video recording of Dr Brian Kennedy, CEO of the Buck Institute for Research on Aging in Novato, California, in conversation with Judith Campisi, 5th February 2013, available at *http://vimeo.com/58981629.*

Chapter 20: The Treatment Revolution

The epigraph comes from Sharon Begley's article 'The Cancer Killer' in *Newsweek*, January 13, 1997.

This chapter relied heavily on the book *p53 in the clinic*, edited by Pierre Hainaut, Magali Olivier and Klas Wiman (Dordrecht, Netherlands: Springer, 2013).

Other important sources were:

p53-based Cancer Therapy by David P. Lane, Chit Fang Cheok and Sonia Lain, in *Cold Spring Harbor Perspectives in Biology*, 2010, 2: a001222. See *http://cshperspectives.cshlp.org*.

'Cancer Specific Viruses and the Development of ONYX-015' by Frank McCormick, in *Cancer Biology & Therapy* 2, Suppl. 1, 2003, 157–160.

'Clinical Trials with Oncolytic Adenovirus in China' by Wang Yu and Hu Fang, in *Current Cancer Drug Targets*, 2007, 7, 659–670.

'MDM2 antagonist Nutlin-3a potentiates antitumour activity of cytotoxic drugs in sarcoma cell lines' by Hege O. Ohnstad *et al.* in *BMC Cancer* 2011, 11, 211.

'An evaluation of small-molecule p53 activators as chemoprotectants ameliorating adverse effects of anticancer drugs in normal cells' by Ingeborg M. M. van Leeuwen *et al.* in *Cell Cycle*, 2012, 11, 1851–61.

'Cyclotherapy: opening a therapeutic window in cancer treatment' by Ingeborg M. M. van Leeuwen, in *Oncotarget*, 2012, 3, 596–600.

Talk by Gerard Evan, entitled 'Cancer isn't mysterious', to staff at Cancer Research UK in September 2012. Available at *www.frequency.com/video/cruk-passion-ta/59881838*.

Acknowledgements

This book might never have got off the ground if the Pathological Society of Great Britain and Ireland had not been prepared to support the proposition that even the toughest science deserves a popular audience. I am extremely grateful for their financial contribution to my research, and for the enthusiasm of the board for the project. In particular I should like to thank Peter Hall, who first suggested this book and who sat with me for a couple of days in his office at Queen's University, Belfast, drawing up a timeline for the history of p53, and Alastair Burt and Simon Herrington, who championed the idea with the Path Soc committee.

My sincere thanks are due also to the following people: Anna Day of Dundee University, who got the project under way; The Society of Authors, whose research bursary enabled me to set off on my travels; my agent, Donald Winchester of Watson Little Ltd, for his quiet encouragement, sure instinct and always good advice; Jim Martin of Bloomsbury Sigma for his enthusiasm and commitment to the story and my editor Caroline Taggart for her keen eyes and good suggestions; Suzanne Cherney, one of the best editors WHO ever had, for her skilful reading of the manuscript and excellent comments (and as much as anything for the long friendship and the laughter); my friend and fellow writer Claire Bell for her support during the writing and wise feedback at a critical moment; my translators in Brazil, Fernanda Paschoal Fortes (who introduced me to *caipirinhas*!) and Henrique Campos Galvão; and Elizabeth Garret for her generous hospitality at the perfect writer's retreat, Cliff Cottage on the ragged Aberdeenshire coast. Special thanks are due also to my partner, Fred Bridgland, whose constant support, listening ear and understanding of a fellow writer's obsession are deeply appreciated.

I am hugely grateful to the many scientists who gave me their time and shared with me something of their personal journeys in science and the highs and lows of their research. Among them I owe a particular debt of gratitude to Pierre Hainaut, without whose advice, guidance and readiness to clarify some extremely complex science from time to time I would have struggled mightily. My understanding of the p53 story was greatly enhanced by conversation and communication with a number of people whose names do not appear in these pages, and to whom I extend my thanks. They include Walter Bodmer, Jean-Christophe Bourdon, Xin Lu, David Meek, Thea Tlsty, Karen Vousden and Geoff Wahl. I am grateful also to the following for their kind permission to use short quotations of theirs from other sources: Suzy Baker, Bill Bryson, Judith Campisi, Richard Lockshin, Matt Ridley and Bob Weinberg.

Finally, I should like to express my special thanks to Luana Locke, John Berkeley and the families affected by Li-Fraumeni syndrome in Brazil whose poignant stories of life in the shadow of cancer shift the focus of attention from the scientists' laboratories to the outside world, and serve to underline just how important p53 research is to us all.

Sue Armstrong, August 2014

Index